# Short Answer Questions For Preclinical Phase Final Exams.

First edition

# Short Answer Questions For Preclinical Phase Final Exams.

**Dr S Steele, Mr J J Harding and Dr S O'Connor**

Academic Medical Press

Published by Academic Medical Press, a division of Academic Medical Consulting.
Nottingham, UK.

Academic Medical Consulting.
Ebury Road, Carrington, Nottingham NG5 1BB
*Somniare audemus*



First published 2014



ISBN 978-0-9566443-7-4

Further copies can be obtained from: http://www.lulu.com
http://www.amazon.com
http://www.amazon.co.uk

# Preface to the first edition

*Short Answer Questions for Preclinical Phase Final Examinations* is intended for medical students approaching major preclinical exams. A frequent complaint from such students is that they do not have realistic high quality exam questions on which to practise. We have tried to address this issue by making use of the experience that Academic Medical Consulting and its staff have accumulated in preparing students for such exams, by writing dedicated revision questions for such individuals.

For this book we have written three complete *mock exams* that cover the full range of preclinical subjects in a typical undergraduate course. Hence subjects as diverse as Physiology, Biochemistry, Psychology and Medical Ethics are represented amongst these questions. Furthermore, the difficulty of these questions is designed to mirror the standard expected by contemporary medical schools. Most importantly we have endeavoured to reflect the current trend amongst medical schools to write *cross-modular* questions that challenge the students to *integrate* the information that they have acquired during their training, by reference to *clinically important scenarios*. This combination of features allows the mock examinations in this text to be a rare opportunity to learn formatively from these summative assessments.

We recommend that the ambitious student begins to practice working through these questions after they have covered the majority of their lectures and are within two months of the relevant exams. In traditional medical schools these questions will be most usefully addressed at the end of the preclinical phase; in more modern schools the equivalent time is 24-30 months into the training programme. Greatest value will be obtained from this text if the questions are attempted under exam conditions.

Nottingham, December 2013 Mssrs Steele, Harding and O'Connor

# CONTENTS

# Examination One

# Test One

***Time allowed - 20 minutes***

1) What type of epithelium lines the lower trachea? (2 marks)

2) Name a major function of type II pneumocytes. (1 mark)

3) What is tidal volume and what is its average volume in an adult? (2 marks)

4) What is the anatomic dead space (serial dead space) and what is its average volume? (2 marks)

5) A chest X-ray shows that a patient has established apical TB. As a result which of the dead spaces will be increased? (1 mark)

6) Name the four drugs usually used in the initial treatment of TB. (4 marks)

7) If the drug treatment is prolonged, the medications are reduced to just two. Which are the two drugs? (2 marks)

8) If the interpretation of the chest X-ray was incorrect and the apical mass is in fact a malignancy, which primary lung cancer is it most likely to be? (1 mark)

9) Name the three components of the levator ani muscle. (3 marks)

10) List the nerve root levels of the pudendal nerve. (2 marks)

**Total 20 marks**

## Test Two

***Time allowed – 20 minutes***

1) A 20 year old male presents with a two day history of colicky loin to groin pain. What is the most likely diagnosis? (1 mark)

2) The General Practitioner, GP, takes a urine dipstick. Name two possible dipstick findings that might also have resulted from the underlying pathology. (2 marks)

3) On taking a more detailed history the GP becomes aware that the patient has a strong family history of renal stones. In addition the patient has had three similar episodes in the last two months. Further biochemical investigations reveal a very unusual composition of the stones. The **K**idney-**U**reters-**B**ladder, KUB, X-ray picture shows poorly defined masses at the ureterovesical junction.

Describe the likely composition of these stones and the reason for the predisposition to renal stones. (3 marks)

4) Name three common complications of urinary tract obstructions. (3 marks)

5) The young man has a sister who suffers from nephrotic syndrome. What is the commonest cause of nephrotic syndrome in childhood? (2 marks)

6) Name four core clinical signs of nephritic syndrome. (4 marks)

7) With respect to health care what is primary prevention? (2 marks)

8) What are the three major sets of factors that influence the uptake of screening services? (3 marks)

**Total 20 marks**

## Test Three

***Time allowed – 20 minutes***

A 15 year old male is admitted urgently with an acute abdomen. The surgical team managing him suspect appendicitis, offer emergency surgery and gain consent from both the parent and the boy. The surgery is successful and a ruptured appendix is excised. Two days after the operation the teenager spikes a fever of 37.9°C. He becomes short of breath and on auscultation bronchial breathing is noted. Unfortunately he develops a severe pneumonia and despite treatment with antibiotics his pneumonia progresses and respiratory failure ensues. After his demise a post mortem shows the lung appearances of Adult Respiratory Distress Syndrome (ARDS). The boy's mother is the city's mayor and she is furious at the death of her son. She accuses the surgeons of negligence.

1) Which three core criteria must be satisfied to meet the legal definition of medical negligence? Explain each of the criteria. (6)

The doctors' legal representative argues that a postoperative pneumonia was a predictable complication of major surgery and that the patient was told of this whilst informed consent was being obtained.

2) Which of the core medical ethics principles is informed consent designed to protect? (1 mark)

The mayor argues that she only agreed to the surgery because of the pressure that she was put under by the doctors and that her son was too young to make such an important decision.

3) In medicolegal terms explain the basis of her challenge. (4 marks)

The doctors' legal representative argues that the mayor's perception of events was affected by the stress of her son's illness. In contrast the mayor's representative argues that the attending surgeon jollied her along, did not listen to what she was saying and continually interrupted her.

4) What is the psychosocial basis of her current position? (2 marks)

5) Name three other behaviours that are in the same category as those that have led to this complaint. (3 marks)

6) The mayor has just experienced one of the top ten most stressful life events. Name four others. (4 marks)

**Total 20 marks**

## Test Four

***Time allowed – 20 minutes***

A politician has been campaigning furiously right up until the night of the elections. In his tired and anxious state he stays up all night to hear the results as they come in. Finally at 6am he receives confirmation that he has won the election. He is elated and expresses himself energetically. Unfortunately as he is jumping for joy he slips, falls and cuts himself on his left forearm. The gash is long and deep enough to require stitches.

As the junior doctor who sees him, you decide to give him a course of antibiotics.

1) Which bacteria are most likely to cause a skin infection here? (1 mark)

The politician is given cefuroxime as antibiotic prophylaxis. Three months later he returns to his GP and complains that he has been experiencing episodes of fever. On examination right upper quadrant tenderness is noted and a history of involuntary weight loss is elicited. The GP suspects a hepatic abscess and refers him to the large local teaching hospital. A diagnostic aspirate is taken and sent to microbiology – this confirms a mixture of bacteria in the abscess. A regime of clindamycin and metronidazole is started.

2) Against what range of bacteria is metronidazole most effective? Give two common examples. (4 marks)

3) The politician does not improve and after a week he becomes septic. He is transferred to the intensive care unit and placed on a respirator. Name two common hospital acquired bacteria that may affect the politician's recovery. (2 marks)

4) Name two commonly used classes of antibiotics that are effective against Pseudomonas aeruginosa and give an example of each. (4 marks)

5) For these antibiotics to be effective they must be transported to the site of action. Define drug bioavailability. (2 marks)

The politician develops a pneumonia. His lung function declines in line with the widespread acute inflammation triggered by the bacteria.

6) Name the dominant white blood cell in acute pneumonia. (1 mark)

7) The pneumonia progresses and the sepsis becomes worse. Name three endogenous pyrogens that could have contributed to his elevated temperature. (3 marks)

8) Despite the antibiotics the patient dies shortly afterwards. At post mortem his lungs are found to be twice their normal weight and it is noted that the pneumonia had not resolved. The pathologist decides that the failure of three organs is likely to have directly contributed to the patient's death. Name the three different organs. (3 marks)

**Total 20 marks**

# Test Five

***Time allowed – 20 minutes***

A 70 year old woman is brought into hospital as an emergency admission. Her daughter believes that her mother fell and struck her head. An examination of the scalp indicates a laceration four centimetres in length overlying the right temple. The mother is unconscious.

1) If she has suffered a haemorrhage name the four spaces or compartments deep to the cranium into which she might have bled. (4 marks)

2) During the history taking you learn from the daughter that the unconscious patient has a medical history of adult polycystic kidney disease. How is adult polycystic kidney disease inherited and which type of haemorrhage do you now favour? (3 marks)

3) Name four clinical manifestations of this type of haemorrhage. (4 marks)

4) Name the two major arteries that directly supply the Circle of Willis. (2 marks)

5) The daughter is upset by the state of her mother but is also anxious because she is eight months pregnant. If she delivers the baby at nine months, what are the definitions of the ranges of normal birth weights, low birth weights, very low birth weights and extremely low birth weights, into which her child might fall? (4 marks)

6) Considering the foetal aetiology of adult disease, low birth weight has been directly linked to several *common* chronic adult illnesses/diseases. Name three. (3 marks)

**Total 20 marks**

## Test Six

***Time allowed – 20 minutes***

A 78 year old male who worked in heavy industry all his life presents with 5 kg of recent weight loss, chest pain that is worsened by breathing and stony dullness of the right lower lung field to percussion. His General Practitioner, GP, is concerned.

1) Which two diagnoses do you think that his GP is most concerned about? (2 marks)

2) What is the descriptive term for the chest pain he is experiencing? (1 mark)

3) A pleural aspirate is taken and sent to the cytopathology department. If the original lesion is pleural in origin, what cell type is likely to be the dominant presence? (1 mark)

4) If the lesion is pleural in origin, state the most likely risk factor, and subdivide this into its two types. (3 marks)

5) Surprisingly, the cytopathology report indicates adenocarcinoma cells. Name four different histological types of malignant lung neoplasms. (4 marks)

6) It is believed that the patient has a pleural effusion. Name four other fluids that can accumulate in the pleural space and indicate the descriptive term for each condition. (4 marks)

7) On the side of the thorax with a pneumothorax, indicate the changes in the following:
*Chest wall movement,*
*Percussion,*
*Breath sounds,*
*Vocal resonance,*
*Mediastinal shift*. (5 marks)

**Total 20 marks**

## Test Seven

***Time allowed – 20 minutes***

A longstanding type 1 diabetic presents with signs and symptoms of renal impairment. He is referred to a specialist nephrologist. The nephrologist takes a 24 hour urine sample and notes a total of 3.5g of protein.

1) What is the diagnosis and what is the likely underlying cause? (2 marks)

2) What are the other major signs and symptoms associated with such a large protein loss in the urine? (3 marks)

3) Name four other conditions/diseases that can cause such a large protein loss. (4 marks)

4) The biopsy of this diabetic's kidney shows malignant lymphocytes. There is a suspicion of lymphoma. What are the seven essential alterations for malignant transformation? (7 marks)

5) Which oncogene overexpression in breast cancer represents an overactive EGF receptor? (2 marks)

6) What is the role of the retinoblastoma protein in oncogenesis? (2 marks)

**Total 20 marks**

# Test Eight

***Time allowed – 20 minutes***

An 80 year old man is admitted to hospital with a history of peripheral vascular disease and hypertension. He appears to have developed new focal neurological signs.

1) "*Acute focal neurological deficit of vascular origin*" refers to which disease or disorder? (1 mark)

2) Name three lifestyle risk factors for a stroke. (3 marks)

3) Name five types of emboli other than thromboemboli. (5 marks)

4) "*Acute focal neurological deficit of vascular origin that lasts less than 24 hours*" refers to what disease or disorder? (1 mark)

5) On taking a detailed history and examination, it becomes clear that *the new neurological signs* of the 80 year old man have disappeared. However he demonstrates stable signs and symptoms that he has had since he was 50 years old. These include a festinant gait, micrographia and a tremor. Indicate which part of the brain has the primary lesion and which neurons are affected. (3 marks)

6) What class of drug is entacapone and how is it able to treat the disease described in (5)? (3 marks)

7) Why is bromocriptine useful in the management of this disease and to what class of drug does bromocriptine belong? (2 marks)

8) Explain why anticholinergics are useful in the management of this disorder and name a drug that is used. (2 marks)

**Total 20 marks**

## Test Nine

***Time allowed – 20 minutes***

A 25 year old woman presents with jaundice to her GP.

1) What type of disease is usually responsible for pre-hepatic jaundice? (1 mark)

2) Outline the faecal and urine colour changes of pre-hepatic jaundice. (2 marks)

3) Name four common signs of hepatic impairment. (4 marks)

4) Although the jaundice resolves, the cause of the jaundice is not identified. The GP is concerned that the underlying disease may progress to cirrhosis, through cycles of relapse and remission. What are the three major complications of cirrhosis? (3 marks)

5) This 25 year old woman is 35 weeks into her pregnancy. What is the normal undistressed foetal heart rate? (2 marks)

6) Name the five classic ways in which presentation of a foetus occurs. (5 marks)

7) What are the two key hormones of lactation? (2 marks)

8) Unfortunately the neonate is born suffering from a nerve palsy that affects the eyes. Which cranial nerve supplies the orbicularis oculis? (1 mark)

**Total 20 marks**

# Test Ten

***Time allowed – 20 minutes***

A 48 year old man with a biopsy confirmed rectal adenocarcinoma undergoes major surgery (an anterior resection). Although the surgery is technically successful the patient has a myocardial infarction.

1) A myocardial infarction under general anaesthesia is not an uncommon event. Why does this occur? (1 mark)

2) Which structural proteins are used as serum diagnostic markers of a myocardial infarction? (2 marks)

3) In a normal cardiac muscle cell under resting conditions to what do Troponin I and Troponin T bind respectively? What is the primary function of the troponin-tropomyosin complex? (3 marks)

4) Despite successful diagnosis and treatment the patient develops an arrhythmia. In a normal heart, starting at the sinoatrial node, describe the idealized pathway of electrical transmission leading to ventricular contraction. (4 marks)

5) Name and briefly describe the four major classes of anti-arrhythmic drugs that affect the depolarization or repolarization phases of the action potential. (4 marks)

6) Which ion channel is responsible for the *absolute refractory period* of cardiac myocytes, preventing the initiation of another action potential? (1 mark)

7) The arrhythmia that the patient develops is atrial fibrillation. What pulse characteristics are typical of atrial fibrillation? (1 mark)

8) Atrial fibrillation has implications for Virchow's triad. Describe Virchow's triad. (3 marks)

9) Name two physiological growth factors that can induce angiogenesis. (1 mark)

**Total 20 marks**

# Examination Two

# Test Eleven

***Time allowed – 20 minutes***

You are a junior doctor working in an oncology department. Your responsibilities include breaking bad news to patients and interpreting and communicating pathology reports.

1) A pathology report on a breast biopsy indicates a local breast cancer. What feature must the histopathologist have seen under the microscope ***to be sure*** that the tumour is a malignancy? (1 mark)

2) A previous CT scan shows that the affected patient has widely disseminated breast cancer. What are the eight essential steps that a malignant cell must undergo to successfully metastasize? (8 marks)

3) What is another word with equivalent meaning to *metastases*? (1 mark)

4) Complete the following table (3 marks):

| **Benign** | **Malignant** |
|---|---|
| Osteoma | ___________ |
| ___________ | Adenocarcinoma |
| Melanocytic naevus | ___________ |

5) What are the four common classical malignancies that have a predominantly osteolytic effect on bone? (4 marks)

6) The patient is believed to carry the aetiological BRCA1 mutation. What class of neoplastic gene is represented by BRCA1? (1 mark)

7) The retinoblastoma gene is in this same functional class and unfortunately the patient is also carrying a mutation in this gene. Under physiological circumstances what is the major transcription factor to which the retinoblastoma protein binds to carry out its function? (1 mark)

8) Eventually the breast cancer progresses to the brain. As this brain cancer is inoperable only supportive management is possible. Dexamethasone is part of the management. To what class of drug does this belong and what syndrome may occur with chronic use? (1 mark)

**Total 20 marks**

# Test Twelve

***Time allowed – 20 minutes***

A 50 year old man with longstanding rheumatoid arthritis visits his GP complaining of shortness of breath on moderate exertion. He appears cyanotic. Auscultation reveals basal crepitations without a wheeze. No stony dullness is found. No pyrexia is noted. The patient does not remember when the shortness of breath started but thinks it has been going on "for quite a while."

1) What is the characteristic histological feature of idiopathic pulmonary fibrosis (cryptogenic fibrosing alveolitis)? (1 mark)

2) Considering its effect on lung function, what class of lung disease is idiopathic pulmonary fibrosis? (1 mark)

3) What are the classic complications of idiopathic pulmonary fibrosis? (4 marks)

4) What type of respiratory failure is characterized by hypoxia and hypercapnia? Name two common causes of this type of respiratory failure. (3 marks)

5) What is the therapy for stage 2 (moderate) COPD? (2 marks)

6) Name four other common causes of right heart failure. (4 marks)

7) One of the pharmacological treatments for heart failure is diuretic administration. What is the pharmacodynamic target of spironolactone and what is the particular clinical significance of this drug? (2 marks)

8) Right heart failure is a cause of oedema. What type of oedema occurs? Name two other major organ impairments that can also cause oedema. (3 marks)

**Total 20 marks**

## Test Thirteen

***Time allowed – 20 minutes***

A 40 year old daughter takes her 70 year old father to their GP, complaining that he has recently become more forgetful.

1) What is dementia? (2 marks)

2) Name the two commonest disease causes of dementia. (2 marks)

3) What is the name of the validated, brief and commonly used clinical test that assesses dementia? What is the maximum possible score on this test? What is the range of scores that is defined as normal? (3 marks)

4) What are the characteristic CNS histological features of Alzheimer's disease? (3 marks)

5) Dementia is a chronic morbidity that is often associated with depression. What are the two commonly used screening questions for depression? (2 marks)

6) Outline six common clinical manifestations of depression that may facilitate diagnosis. (6 marks)

7) Name two commonly used psychological interventions in the management of depression. (2 marks)

**Total 20 marks**

# Test Fourteen

***Time allowed – 20 minutes***

1) Which major nerve lies deep to the bicipital aponeurosis in the antecubital fossa? (1 mark)

2) From which roots in the brachial plexus does the median nerve originate? (2 marks)

3) What is the name of the branch of the median nerve that provides the motor supply to the thenar muscles? (1 mark)

4) Name four common causes of carpal tunnel syndrome. (4 marks)

5) Name four signs or symptoms that are commonly presented by a sufferer of carpal tunnel syndrome. (4 marks)

6) Name three tests used to diagnose carpal tunnel syndrome. (3 marks)

7) Describe the mechanism of action of hydrocortisone as an immunosuppressive and anti-inflammatory drug. (5 marks)

**Total 20 marks**

## Test Fifteen

***Time allowed – 20 minutes***

A 75 year old man presents to his GP complaining of pain on micturition. He has a history of recent urinary tract infections.

1) Why is this man likely to be predisposed to urinary tract infections? (2 marks)

2) What is the most likely source of the infecting bacteria and what type of bacteria are they likely to be? Name the most likely bacterium. (3 marks)

3) Unfortunately a prostatic biopsy indicates that the patient has a small peripheral focus of malignant tumour. Name the most probable histological type of malignancy. (2 marks)

4) What is the commonest grading system used for this malignancy? (1 mark)

5) The patient complains of chronic loin pain, fever and lethargy. The GP notices that the patient is now hypertensive. What is the most probable diagnosis and most worrying complication? (2 marks)

6) Name three commonly used antibiotics in the treatment of urinary tract infections. (3 marks)

7) Name four other species of bacteria that commonly cause UTIs and indicate the commonest infecting scenario for each (4 marks):

8) Describe the mechanism of action of trimethoprim. (3 marks)

**Total 20 marks**

## Test Sixteen

***Time allowed – 20 minutes***

A 37 year old man is an emergency admission to hospital with suspected meningitis.

1) Name the three layers of the meninges. (3 marks)

2) Assuming that this is a bacterial meningitis, which two bacteria are the most probable aetiological agents? (2 marks)

Although the patient has some meningeal signs he starts to display odd behaviour and a lowered level of consciousness. You believe that he may be suffering from encephalitis.

3) What type of encephalitis would be of greatest concern in the acute setting and why? (2 marks)

4) What is the mechanism of action of the leading drug used in the treatment of this type of encephalitis? (2 marks)

5) Complete the following table (8 marks):

*Summary of typical CSF findings in meningitis/encephalitis:*

| | Normal CSF | Bacterial Infection | Viral Infection |
|---|---|---|---|
| Cells | 0-5 wbc/ml | >1000 wbc/ml | <1000 wbc/ml |
| Polymorphs | 0 | | |
| Lymphocytes | 5 | | |
| Glucose | 40-80mg/dl | | |
| (CSF:plasma) glucose ratio | 66% | | Normal |
| Protein | 5-40 mg/dl | | Negligible change |
| Culture | Negative | Positive | Negative |
| Gram staining | Negative | Positive | Negative |

6) If the treatment of the encephalitis (or meningitis) is too late or insufficient, raised intracranial pressure may occur. This can lead to fatal herniation. One of the subtypes of herniation is transtentorial (central). Name the structure over which this herniation occurs and indicate the layer of the meninges from which it is derived. (3 marks)

**Total 20 marks**

# Test Seventeen

***Time allowed – 20 minutes***

Alcohol abuse has a range of deleterious short term and long term effects. As a result, assessing alcohol abusers is an important clinical task.

1) What four screening questions are commonly used to identify alcohol abusers? (4 marks)

2) What are the recommended safe maximum number of units of alcohol that are permitted for men and women in daily and weekly terms? (4 marks)

3) Name five gastrointestinal malignancies at a greater risk of occurring in alcohol abusers. (5 marks)

As a junior doctor on an oncology ward you have to manage terminally ill patients.

4) Define the doctrine of double effect as it applies to clinical practice. (3 marks)

5) Classically under what moral conditions is the doctrine of double effect applicable? (4 marks)

**Total 20 marks**

## Test Eighteen

***Time allowed – 20 minutes***

1) Define disease incidence. (1 mark)

2) When assessing the health of a population, it is often necessary to rely on surveys of patients. What are the two biggest issues regarding the accuracy of health information acquired in this way? (2 marks)

3) Define attributable risk. (2 marks)

4) List the Bradford Hill's criteria for inferring causality (9 marks).

5) Explain what is meant by the *general fertility rate*. (2 marks)

6) Explain what is meant by the phrase "population census." (2 marks)

**7)** What is the WHO definition of health? (2 marks)

**Total 20 marks**

# Test Nineteen

***Time allowed – 20 minutes***

A teenager presents at his GP surgery complaining of new onset shortness of breath on exertion with an obvious wheeze.

1) Name three common atopic disorders. (3 marks)

2) List four histological features of asthma. (4 marks)

3) What functional class of lung disease is asthma? (1 mark)

4) The young boy has just started smoking. If he continues to smoke what non-neoplastic lung diseases will he be prone to? (3 marks)

5) Which of these diseases (question 4) is most likely to be exacerbated by alpha-1 antitrypsin deficiency? (1 mark)

6) Chronically what effect will alpha-1-antitrypsin deficiency ultimately have on the liver? (2 marks)

7) What colour changes will occur in the faeces and urine of such an individual? (2 marks)

8) Haemochromatosis and Wilson's disease can ultimately cause similar changes in the liver. What are the underlying disease processes in each case? (4 marks)

**Total 20 marks**

## Test Twenty

***Time allowed – 20 minutes***

A 47 year old amateur sportsman was admitted to his local Accident and Emergency department after complaining of chest pain. It is Sunday night and he spent the afternoon playing squash. The previous day he played football for 3 hours. The chest pain is described as 7/10 in intensity, central and radiates to his neck. It started two hours after he finished his game of squash, and is unusual in not being relieved by ibuprofen. The pain has continued and is still present on admission.

He has no family or personal history of arrhythmia, no evidence of hypertrophic obstructive cardiomyopathy (HOCUM or HOCM) and no evidence of hyperlipidaemia. He has never smoked. He has no personal or family history of diabetes mellitus. His parents are in their 70s and are in good health – neither has suffered from angina or a myocardial infarction.

Palpation on the ribs or sternum could not elicit the pain.

The junior doctor managing the patient decides that the most likely diagnosis is a myocardial infarction.

1) Name four drugs routinely used in the immediate management of a myocardial infarction. (4 marks)

2) Clopidogrel and enoxaparin share a major side-effect (adverse effect). Name this adverse effect. (1 mark)

3) Describe the mechanism of action of enoxaparin. (3 marks)

The junior doctor requests an ECG and takes blood to test for the release of cardiac enzymes/troponin. The ECG changes are indeterminant with no clear ST elevation. However, troponin and CK-MB levels are significantly raised in the venous blood.

4) Why would a myocardial infarction raise circulating levels of troponin and CK-MB? (1 mark)

5) Although the patient finds the drug regime effective, he complains that one of the medications is giving him a headache. Which medicine is it likely to be and why is the headache occurring? (2 marks)

6) How and why are streptokinase and tissue plasminogen activator, tPA, useful in the management of a myocardial infarction? (4 marks)

Three days after admission the patient is still suffering from a similar pain and has developed a pyrexia of 38°C. Angiography indicates that the three major coronary arteries are each more than 90% patent. The relatively young age and paucity of risk factors caused the consultant supervising the junior doctor to reconsider the working diagnosis.

7) What is your favoured new diagnosis? (2 marks)

8) What medications would you use to manage the latest diagnosis? (3 marks)

**Total 20 marks**

# Examination Three

# Test Twenty-One

***Time allowed – 20 minutes***

It is 11.30pm on a Monday night in a delivery suite at the local teaching hospital. A husband is holding his wife's hand as she enters the 22nd hour of labour. Despite the mother's small to medium build, the child appears large and this seems to be making the delivery difficult. The midwife has a student nurse assisting in the suite. Being an exceptionally well trained and gifted student she remembers details of her *introduction to biochemistry* course.

1) On the course she remembers being told a word for *big babies*. What is the word most likely to have been? (1 mark)

2) The nursing student remembers that there was a common endocrinological cause for such large babies – gestational hyperglycaemia. Name four hyperglycaemic hormones. (4 marks)

3) Accordingly, to what disorder is such a pregnant woman prone? (1 mark)

During the long labour the mother-to-be has plenty of time to think. Her mind drifts to her 22 year old male cousin. She was close to her cousin and had been quite distressed when six weeks ago he lost a lot of weight very quickly, started to drink a lot of fluids and was repeatedly going to the toilet. She noticed that her cousin's breath also smelt funny and his eyes looked sunken. Eventually, after being persuaded to visit his GP, he was rushed into hospital.

4) What is the name for the condition that the young man was in and which protein was part of the core treatment? (2 marks)

5) Why had he lost weight and developed sunken eyes? (2 marks)

6) Name the cause of the smell and give two examples of substances that could have been responsible for the smell. (3 marks)

7) What disease is her cousin likely to be manifesting? (2 marks)

8) i) Is 12 mM a normal physiological concentration of blood glucose?
ii) Is 4.2 mM a normal physiological concentration of blood glucose?
iii) What initial effect does insulin have on the plasma potassium concentration?
iv) How does insulin have this effect on plasma potassium concentration?

(5 marks)

**Total 20 marks**

# Test Twenty-Two

***Time allowed – 20 minutes***

It is 8pm and you are a junior doctor called to see a 33 year old female patient who has a bad headache with associated nausea, vomiting and photophobia. She has been an inpatient for two weeks for the management of recurrent ulcerative colitis. She had been able to eat a lunch that included chocolate cake and lemon sorbet. The headache has continued for eight hours despite her nurse giving her two 500mg paracetamol (acetaminophen) tablets. Her blood pressure and pulse are normal. On examination you are unable to elicit signs of meningism and no fever is noted.

1) Based on this information what is the most likely diagnosis? (1 mark)

2) The most likely diagnosis can classically be divided into four phases. Name the four phases. (4 marks)

3) In general terms what pathophysiological process in the brain is directly associated with the aura? (2 marks)

4) Name three common different types of headache. (3 marks)

5) Which part of the brain is believed to be the migraine generator? (1 mark)

6) It is believed that the pathophysiology of this severe headache involves inflammatory neuropeptides. Based upon this information what class of drug would you use to treat this headache? (2 marks)

7) The most likely diagnosis of the headache is also associated with epilepsy. Name four established causes of epilepsy. (4 marks)

8) The unfortunate patient goes on to develop epilepsy that is recurrent and resistant to treatment. Seizures occur in the daytime and affect the level of consciousness. (a) What ethical principle governs your reporting of this patient to the governmental authorities responsible for road safety? (b) What duty must you contravene under such circumstances? (2 marks)

9) If doctors report their patients' misdeeds to the authorities they may cease to be trusted by their patients. Patients will be less willing to seek out doctors and trust them – damaging the

doctor-patient relationship. If the doctor bases his decision not to report the epileptic patient on the likely subsequent effects, what ethical morality/view is being demonstrated? (1 mark)

**Total 20 marks**

## Test Twenty-Three

***Time allowed – 20 minutes***

An enthusiastic weekend cyclist attends visits his general practitioner to complain about wrist pain and numbness in the lateral two digits of the affected hand whilst he is cycling. He uses drop-down handlebars and usually cycles for at least 90 minutes continuously during each session. This 37 year old man is otherwise fit and well with no significant medical history.

1) Which cords give rise to the median nerve? (2 marks)

2) Name the five terminal branches of the brachial plexus. (2.5 marks)

3) Name the intrinsic muscles of the hand supplied by the median nerve. (2.5 marks)

4) Why do you think that the cyclist is experiencing this pain? Briefly explain the pathogenesis. (2 marks)

5) The cyclist has positive results from Tinel's test, Phalen's test and Reverse Phalen's test.

From which syndrome is the cyclist likely to be suffering? (1 mark)

6) Name four classic causes of this syndrome. (4 marks)

7) The cyclist has been taking an NSAID for his wrist pain. Inhibition of which gastric enzyme is likely to be the cause of any subsequent chronic gastric pain? (2 marks)

8) Two years later the cyclist is diagnosed as having rheumatoid arthritis. What is the underlying aetiology of rheumatoid disease? (4 marks)

**Total 20 marks**

# Test Twenty-Four

***Time allowed – 20 minutes***

An 18 year old boy presents to his GP complaining of pain on urination. After a complete history and examination it becomes apparent the young man has been suffering from a urethral discharge for two days.

1) How is he most likely to have acquired this urethritis and what are the two commonest causes? (3 marks)

2) The patient has a 30 year old female partner, Sarah, who is informed of his condition. Six months later she is diagnosed with squamous cell carcinoma in situ of the cervix (CIN 3). She is distraught and blames her partner's E.coli urethritis for the neoplastic diagnosis. Is Sarah correct? Explain. (4 marks)

3) Despite treatment she goes on to develop a cervical cancer. Cervical cancers are common in the general population. Name three different cervical malignancies. (3 marks)

Three months after being diagnosed with cervical cancer and receiving treatment she visits her GP complaining of chest pain. The chest pain was sharp and associated with shortness of breath for two days. There is no radiation, nausea or swelling of ankles. She had been apyrexial.
The chest pain was worsened by the movements of breathing.

4) What is the most likely diagnosis and what drug would you select for initial treatment? (2 marks)

5) List three possible ECG changes that can occur with the above diagnosis. (3 marks)

The young man with urethritis, Peter, breaks up with his ailing and cancer ridden partner. He is persuaded by his parents to spend some time with his uncle, George, on his farm whilst he sorts out his life. Peter initially does well, appearing to adapt to his new location, making friends and finding work at the local newspaper. After about a year Peter's uncle starts to notice some odd behavioural changes. Peter begins to spend more time by himself, he sporadically fails to attend work without any clear reason and pays noticeably less attention to his personal hygiene. His uncle takes him to the local GP who gives Peter a short course of Prozac. However, over the next year Peter inexorably declines, stops going to work altogether and avoids his friends. George finds Peter harder and harder to understand in conversation. Peter claims that to have survived his break up shows that he has god like strength of mind and so he must be a god. On Sunday evenings, their weekly game of chess ends unsatisfactorily with Peter claiming that he has discovered a new way for the rook to move.

6) What is the most likely diagnosis of Peter's condition? (1 mark) Identify four supporting clinical features. (4 marks)

**Total 20 marks**

## Test Twenty-Five

***Time allowed – 20 minutes***

A 30 year old man attends for a consultation stating that six months ago he tried but failed to quit smoking. However, he now wants to try again and has already cut down the number of cigarettes he smokes each day. He wants your help and advice to successfully quit smoking.

1) List and briefly describe four of the ICD-10 diagnostic criteria for substance dependence. (8 marks)

2) In the **transtheoretical model of behaviour change**, smokers can be assigned to one of five stages of change. A) At what stage of change is this patient? B) State two items in the question stem that support your answer. (3 marks)

3) Besides the stage of change identified in the previous question, name four other stages of change. (4 marks)

4) List and briefly describe the "5 As" that many clinicians use to guide their delivery of smoking cessation advice. (5 marks)

**Total 20 marks**

## Test Twenty-Six

***Time allowed – 20 minutes***

As a conscientious student you have worked so hard on your undergraduate biochemistry that you start to dream about it in your sleep. In fact in your dreams you are a crime scene investigator, trying to help track down and capture a particularly ruthless and brutal serial killer. Like all good crime scene investigators you have 100% recall of everything you have ever been taught and have a tendency to speak in taut acronym-laden bursts ……

Your serial killer has become overconfident and at the latest crime scene you identify some drops of the killer's blood.

1) What is the name for the simplest classification of blood group antigens? (1 mark)

2) Your blood tests indicate that the killer has the most rare blood type. Which blood type is this? (1 mark)

3) A simple test carried out on the blood displays the chromosomes. What is the name of this display? (1 mark)

4) The test confirms that the serial killer is male. Further tests indicate that your criminal is carrying the genetics (triplet repeat expansions) for a rare neurodegenerative disorder. This disorder is expressed in every generation and its age of onset is earlier and more severe with each generation.

i) In terms of simple Mendelian genetics what type of inheritance is this disease likely to show? (2 marks)
ii) What is the term for the fact that the disease occurs earlier with each generation and is

likely to be more severe? (1 mark)
iii) What is this disease called? (1 mark)

You suspect that part of the killer's motivation may be fatalistic appreciation of his own disease and disease prognosis.

On a thorough examination of the genome it is noted that this extremely unfortunate serial killer is a carrier for the sickle cell disease (he is a heterozygote and so is asymptomatic).

5) Based on this information can you speculate regarding the likely ethnic group of the killer? (1 mark)

You pass on this wealth of information to the police, who are enormously grateful and make you their CSI of the week.

6) What is the nature of the DNA disorder in sickle cell disease? (2 marks)

7) Considering how damaging the effects of having sickle cell disease are, it is surprising that it has not been eliminated as a result of negative evolutionary pressures. Can you suggest a reason why the sickle cell trait (sickle cell carriers) still exists in the population? (2 marks)

8) Because of this selection pressure, Hardy-Weinberg population genetics do not apply well to sickle cell population genetics. What are the assumptions of Hardy-Weinberg population genetics? (5 marks)

9) State the two Hardy-Weinberg equilibrium equations (or expressions). (3 marks)

**Total 20 marks**

## Test Twenty-Seven

***Time allowed – 20 minutes***

A 57 year old female politician attends her GP surgery to discuss her recovery from a recent myocardial infarction. Because she is a driven and intelligent person she has done some internet research on this issue and as a result she has a lot of questions. You feel under pressure to maintain your professional credibility by answering as clearly and as accurately as you can………

1) She has noticed that many of the complications of a myocardial infarction involve either arrhythmias or heart failure (including cardiogenic shock). You agree, but feel duty bound to alert her to the wider range of complications. Name five other classical complications of a myocardial infarction. (5 marks)

2) Your patient remembers reading about an autoimmune complication of a myocardial infarction but cannot remember the name. What is the name of this complication? (1 mark)

3) Name two core signs or symptoms of Dressler's syndrome. Briefly indicate the underlying pathogenesis of each. (4 marks)

4) Name two other autoimmune diseases that directly target the cardiovascular system. (2 marks)

5) Indicate two classical long term complications that can occur as a result of any autoimmune disease. (2 marks)

6) The politician has a sister who suffered from a myocardial infarction at 55 years of age. Assuming that this myocardial infarction occurred as the result of a thrombosis, where in the vasculature is this thrombosis most likely to have occurred? (2 marks)

7) Assuming that this myocardial infarction caused ST elevation, which ECG leads are most likely to reveal this change? (2 marks)

8) Percutaneous coronary intervention was not successful and so thrombolysis was used in the management of this myocardial infarction. Name two such commonly used fibrinolytic agents. (2 marks)

**Total 20 marks**

## Test Twenty-Eight

***Time allowed – 20 minutes***

A 27 year old bicycle courier presents to you as an emergency admission. He complains of leg pain after being driven into by a Lexus. On examination his lower leg is lacerated, swollen and deformed with loss of the ability to weight bear. The courier is in a great deal of pain and you immediately prescribe oral analgesics and request X-rays of the leg. The X-rays show a mid tibia-fibula fracture. The leg is placed in a cast.

1) You suspect that the muscles of the lateral compartment have been crushed and their function impaired. Name the two muscles of the lateral compartment of the lower leg. (2 marks)

2) What are the normal actions of theses muscles? (2 marks)

Three days later the patient returns to hospital complaining of exquisite tenderness of that leg. The pain keeps him awake at night. Because the pain is out of proportion with the presumed diagnosis, acute compartment syndrome is suspected.

3) What are the generally accepted 5Ps of compartment syndrome? (5 marks)

4) A senior doctor mentions that there is a sixth P that can also be useful in the diagnosis of compartment syndrome. Explain to what this sixth P may refer. (2 marks)

5) As well as the muscle weakness and continuing muscle pain the courier mentions that he also has dark urine. As a result there is a suspicion of impairment of which major intrabdominal organ(s)? (1 mark)

6) Initially which blood tests would you request to confirm the clinical suspicion? (3 marks)

7) A blockage can occur in any of the tubular components of the nephron. Name the tubule components of the nephron through which urine is formed and passes. (5 marks)

**Total 20 marks**

## Test Twenty-Nine

***Time allowed – 20 minutes***

A previously well 15 year old boy presents with abdominal pain that is central abdominal, 7/10 in intensity with a 24 hour history. On examination he has no abdominal scars. Two hours after being admitted into hospital he complains that the pain is worse and is now in the right lower quadrant. He spikes a fever of 38.5°C.

1) What is the most likely diagnosis? (1 mark)

2) Name three endogenous chemical factors that may be causing the pyrexia. (3 marks)

3) What is the definitive management of acute appendicitis? (2 marks)

4) Describe the most likely macroscopic appearance of the excised appendix. (3 marks)

5) On examining the inflamed appendix the histopathologist notes an infiltration of numerous cells with multilobed nuclei. What is this dominant cell type likely to be and is the preponderance of this cell type consistent with your favoured diagnosis? (2 marks)

6) Unfortunately for the patient, on thorough examination of the appendix the histopathologist finds an area that amounts to a carcinoma. What are the key cellular features of malignant cells? (4 marks)

7) It takes the teenager a surprisingly long time to recover from the surgery. The surgeon notes that the wound takes longer than usual to heal. Name five **general** systemic factors that can decrease the rate of wound healing. (5 marks)

**Total 20 marks**

## Test Thirty

***Time allowed – 20 minutes***

As a junior doctor working for surgeons, you have to manage many postoperative patients. It is not unusual for one of your patients to be pyrexial.

1) Name four classic (diagnostic) causes of postoperative fever. (4 marks)

2) What is the first line drug used in the treatment of a urinary tract infection (UTI)? (1 mark)

3) What are the first line drugs used in the treatment of deep venous thrombosis (DVT)? (2 marks)

4) Name the tests used to show that heparin is within its therapeutic range and that warfarin is within its therapeutic range. (2 marks)

A 55 year old patient has undergone a heart valve replacement with an artificial valve. As a result she is required to maintain lifelong anticoagulation with warfarin. The aetiology of her heart valve dysfunction is believed to be childhood rheumatic fever. For the first year after the operation the warfarin mediated control of INR is routine. However, after the second year progressively larger doses of warfarin are required to maintain the target INR.

During the third year the patient's renal function and cardiac function remain good, however, the hepatic function deteriorates insidiously. The patient reports that she was divorced by her husband shortly after the valve transplant and is currently dating again. When she puts down her handbag the physician hears the clinking of glass bottles.

5) What is the most likely explanation for deterioration in the control of her INR? (3 marks)

6) If the warfarin levels are maintained at too high a level, to what types of complications will the patient be prone? (1 mark)

7) Over the course of the next five years, she develops spider naevi, caput medusae and haemorrhoids. Blood tests show markedly raised AST, ALT and ALP. Concurrently her personal and social life leave her unhappy and she admits that drinking excessive alcohol is a problem. What affective disorder is chronic alcohol abuse most likely to cause? (1 mark)

8) The GP is concerned about the progression of her psychiatric disorder and decides that she needs pharmacological support. He considers using an SSRI, SNRI or MAOI medications.
Give an example of each of these classes and briefly describe the primary mechanism of action. (6 marks)

**Total 20 marks**

# Answers

## Test One

1) What type of epithelium lines the lower trachea? (2 marks)

**Ciliated pseudostratified columnar epithelium.** (2 marks) This is also more simply termed *respiratory epithelium.*

> This type of epithelium lines the respiratory tract and so is found in the trachea and bronchi. Cigarette smoke irritates this epithelium and causes squamous metaplasia. Unfortunately this causes the loss of cilia and hence the breakdown of the mucociliary escalator.

2) Name a major function of type II pneumocytes. (1 mark)

These pneumocytes have two equally important functions, **surfactant production** and **regeneration of pneumocytes**.

3) What is tidal volume and what is its average volume in an adult? (2 marks)

The tidal volume is the volume of air taken in and out with each **normal breath. This is approximately 500mls (0.5L).**

4) What is the anatomic dead space (serial dead space) and what is its average volume? (2 marks)

The anatomic dead space is part of the **conducting airway that does not take part in gaseous exchange**. Hence the mouth and trachea are part of the anatomic dead space. The anatomic dead space has an average volume of **150mls (0.15L) in an adult**.

5) A chest X-ray shows that a patient has established apical TB. As a result which of the dead spaces will be increased? (1 mark)

**Distributive dead space** (because alveoli will be reduced in number). Additionally **physiological dead space** will be increased because physiological dead space is the sum of distributive and serial dead space. (1 mark for either answer).

6) Name the four drugs usually used in the initial treatment of TB. (4 marks)

**Rifampicin, Isoniazid, Pyrazinamide and Ethambutol.**

Some protocols use *streptomycin* in place of ethambutol as a second-line TB drug.

7) If the drug treatment is prolonged, the medications are reduced to just two. Which are these two drugs? (2 marks)

**Rifampicin.**
**Isoniazid.**

8) If the interpretation of the chest X-ray was incorrect and the apical mass is in fact a malignancy, which primary lung cancer is it most likely to be? (1 mark)

**Adenocarcinoma (periphery of the lung field).**

As a rule of thumb, adenocarcinomas are more common in the periphery of the lung fields and squamous cell carcinomas are more common in the hila.

9) Name the three components of the levator ani muscle. (3 marks)

**Iliococcygeus.**
**Pubococcygeus.**
**Puborectalis.**

10) List the nerve root levels of the pudendal nerve. (2 marks)

**S2, S3 and S4** (two marks for all three correct, otherwise only one mark).

Remember that S2, 3 and 4 keep the pelvis off the floor..........

**Total 20 marks**

## Test Two

1) A 20 year old male presents with a two day history of colicky loin to groin pain. What is the most likely diagnosis? (1 mark)

**Renal stone/Nephrolithiasis.**

Colicky pain implies a partial or complete obstruction of a hollow viscus, with a reactive *intermittent spasm* of the muscle of the viscus wall. In this case the viscus is the ureter and it is being obstructed by a kidney stone. Nephrolithiasis is the more formal term for renal stones.

2) The General Practitioner, GP, takes a urine dipstick. Name two possible dipstick findings that might also have resulted from the underlying pathology. (2 marks)

Any two of:
**Haematuria**. This is an irritative sign that can be caused by a renal stone or an infection.
**Nitrites**. The concentration of nitrites is raised during a urinary tract infection (UTI). The bacteria can convert nitrates to nitrites. Urinary tract infections can themselves predispose to renal stones.
**Raised white cell count.** The bacterial infection triggers a **neutrophilic** response as part of the acute inflammation.

Urinary tract obstruction can cause stasis and predispose to infection.

3) On taking a more detailed history the GP becomes aware that the patient has a strong family history of renal stones. In addition the patient has had three similar episodes in the last two months. Further biochemical investigations reveal a very unusual composition of the stones. The **K**idney-**U**reters-**B**ladder, KUB, X-ray picture shows poorly defined masses at the ureterovesical junction.

Describe the likely composition of these stones and the reason for the predisposition to renal stones. (3 marks)

These are **cystine stones**. The patient may have a family history of **cystinuria** – he has a **defect in the renal transporter that reabsorbs cystine**.

Cystine stones are less common than the more prevalent calcium stones. Because cystine stones have a lower radiodensity than calcium stones they can be more difficult to discern radiologically than calcium stones. Cystinuria tends to have a strong family history that is inherited in an autosomal recessive manner.

4) Name three common complications of urinary tract obstructions. (3 marks)

Any three of:
**Bladder dilation**
**Megaureter**
**Hydronephrosis**
**Acute pyelonephritis**
**Acute renal failure**
**Chronic renal failure**
**Urinary tract infections/sepsis/stone formation**

**Fetal lung hypoplasia**

5) The young man has a sister who suffers from nephrotic syndrome. What is the commonest cause of nephrotic syndrome in childhood? (2 marks)

**Minimal change nephropathy (synonymous with *minimal change glomerulonephritis* and *minimal change disease*).**

> Minimal change nephropathy was named in part because under the light microscope no pathology can usually be identified. The hallmark abnormality of *podocyte fusion* can only be identified under the electron microscope.

6) Name four core clinical signs of nephritic syndrome. (4 marks)

Any four of:
**Haematuria**
**Proteinuria**
**Hypertension**
**Oedema**
**Oliguria**
**Uraemia** +/- pain

7) With respect to health care, what is primary prevention? (2 marks)

Primary prevention refers to health care initiatives aimed at ***maintaining or improving health*** among people who are ***currently free of symptoms.*** This includes strategies such as the modification of risk factors, e.g. smoking, diet and alcohol intake.

8) What are the three major sets of factors that influence the uptake of screening services? (3 marks)

The simplest encompassing classification is:
**Patient factors.**
**Provider factors.**
**Organisational factors.**

**Total 20 marks**

## Test Three

A 15 year old male is admitted urgently with an acute abdomen. The surgical team managing him suspect appendicitis, offer emergency surgery and gain consent from both the parent and the boy. The surgery is successful and a ruptured appendix is excised. Two days after the operation the teenager spikes a fever of 37.9°C. He becomes short of breath and on auscultation bronchial breathing is noted. Unfortunately he develops a severe pneumonia and despite treatment with antibiotics his pneumonia progresses and respiratory failure ensues. After his demise a post mortem shows the lung appearances of Adult Respiratory Distress Syndrome (ARDS). The boy's mother is the city's mayor and she is furious at the death of her son. She accuses the surgeons of negligence.

1) Which three core criteria must be satisfied to meet the legal definition of medical negligence? Explain each of the criteria. (6 marks)

**Duty of care** (1 mark)
"A legal obligation imposed on an individual requiring that they adhere to a standard of reasonable care while performing any acts that could foreseeably harm others." Medical negligence requires a failure of such duty of care. (1 mark)

**Breach of duty** (1 mark)
"The claimant *must show* that the defendant fell below the required standard of care." (1 mark)

**Causation** (1 mark)
"The claimant must establish that his *condition was worsened* or his unimproved condition was caused by the doctor's negligence." (1 mark)

The doctors' legal representative argues that a postoperative pneumonia was a predictable complication of major surgery and that the patient was told of this whilst informed consent was being obtained.

2) Which of the core medical ethics principles is informed consent designed to protect? (1 mark)

**Autonomy.**

> Remember that the four key principles of medical ethics developed by Beauchamp and Childress were: Autonomy, Justice, Beneficence and Non-maleficence.

The mayor argues that she only agreed to the surgery because of the pressure that she was put under by the doctors and that her son was too young to make such an important decision.

3) In medicolegal terms explain the basis of her challenge. (4 marks)

In effect the mayor is claiming that her consent was **coerced**. Her consent was not truly voluntary and so was **not valid informed consent.**
Her son was not **Gillick competent** – he did not have the **capacity to consent at his age**.

(Valid consent must be **voluntary**, be obtained from an **individual with capacity** understand and be **appropriately informed**).

The doctors' legal representative argues that the mayor's perception of events was affected by the stress of her son's illness. In contrast the mayor's representative argues that the attending surgeon jollied her along, did not listen to what she was saying and continually interrupted her.

4) What is the psychosocial basis of her current position? (2 marks)
Poor ***communication*** or poor ***behaviour*** by the surgeon. These are behaviours that tend to **block patient disclosure**.

5) Name three other behaviours that are in the same category as those that have led to this complaint. (3 marks)

Any three of the highlighted list below:
Not listening or interrupting.
**Depersonalization.**
**Explaining away distress as normal.**
**Attending to the physical aspects only.**
Jollying patients along.
**Use of jargon.**

6) The mayor has just experienced one of the top ten most stressful life events. Name four others. (4 marks)

Any four of :

| | |
|---|---|
| 1. | Death of a spouse |
| 2. | Divorce |
| 3. | Marital separation |

4. Jail term
5. Personal injury or illness
6. Marriage
7. Fired from work
8. Marital reconciliation
9. Retirement

(The tenth item is the death of a close family member, such as occurred here).

**Total 20 marks**

## Test Four

A politician has been campaigning furiously right up until the night of the elections. In his tired and anxious state he stays up all night to hear the results as they come in. Finally at 6am he receives confirmation that he has won the election. He is elated and expresses himself energetically. Unfortunately as he is jumping for joy he slips, falls and cuts himself on his left forearm. The gash is long and deep enough to require stitches.

As the junior doctor who sees him, you decide to give him a course of antibiotics.

1) Which bacteria are most likely to cause a skin infection here? (1 mark)

**Staphylococcus aureus** is present on the skin and is highly pathogenic.

> Streptococci are the second most common cause of cellulitis. Part of the concern regarding staphylococcal infection is the possibility of the development of a harmful pathogenic MRSA strain (methicillin resistant staphylococcus aureus).

The politician is given cefuroxime as antibiotic prophylaxis. Three months later he returns to his GP and complains that he has been experiencing episodes of fever. On examination right upper quadrant tenderness is noted and a history of involuntary weight loss is elicited. The GP suspects a hepatic abscess and refers him to the large local teaching hospital. A diagnostic aspirate is taken and sent to microbiology – this confirms a mixture of bacteria in the abscess. A regime of clindamycin and metronidazole is started.

2) Against what range of bacteria is metronidazole most effective in humans? Give two common examples. (4 marks)

Metronidazole is most effective against **obligate anaerobes** (Gram positive and Gram negative). ***Bacteroides*** and ***Clostridia*** species are common in the gastrointestinal tract.

Additionally, metronidazole is effective against anaerobic streptococci that are also present in the gastrointestinal tract. (1 mark for each highlighted word).

3) The politician does not improve and after a week he becomes septic. He is transferred to the intensive care unit and placed on a respirator. Name two common hospital acquired bacteria that may affect the politician's recovery. (2 marks)

Any two of:
**Pseudomonas aeruginosa** (ventilatory support) - 1 mark
**Clostridium difficile** (opportunistic infection post antibiotic therapy) - 1 mark
**Methicillin resistant staphylococcus aureus** (from staff or fellow patients) – 1 mark

4) Name two commonly used classes of antibiotics that are effective against Pseudomonas aeruginosa and give an example of each. (4 marks)

(1 mark for the correct class and 1 mark for each example).
**Carbapenems e.g. imipenem, meropenem and ertapenem**
**Quinolones e.g. ciprofloxacin, levofloxacin and moxifloxacin**

5) For these antibiotics to be effective they must be transported to the site of action. Define drug bioavailability. (2 marks)

The **proportion** of administered drug that reaches the **systemic circulation** and is thus available for distribution to the site of action.

> An orally administered drug is considered to have entered the systemic venous circulation after it passes through the liver and enters the hepatic vein.

The politician develops a pneumonia. His lung function declines in line with the widespread acute inflammation triggered by the bacteria.

6) Name the dominant white blood cell in acute pneumonia. (1 mark)
**Neutrophil.**

> Neutrophils are important in mediating and coordinating acute inflammation. Neutrophils, basophils and eosinophils are termed ***polymorphs*** or ***polymorphonuclear leukocytes*** or ***polymorphonuclear*** (PMN) cells because they each have multilobed nuclei.

7) The pneumonia progresses and the sepsis becomes worse. Name three endogenous pyrogens that could have contributed to his elevated temperature. (3 marks)

Any three of:
**IL-1, IL-2, IL-6, TNF-α, Prostaglandins.**

IL-1 is *Interleukin 1*
IL-2 is *Interleukin 2*
IL-6 is *Interleukin 6*
TNF-α is *Tissue Necrosis Factor-Alpha*

8) Despite the antibiotics the patient dies shortly afterwards. At post mortem his lungs are found to be twice their normal weight and it is noted that the pneumonia had not resolved. The pathologist decides that the failure of three organs is likely to have directly contributed to the patient's death. Name the three different organs. (3 marks)

**Lungs** (pneumonia and possible adult or acute respiratory distress syndrome, ARDS).
**Kidneys** (septic shock potentially causing renal failure and possible multiorgan failure).
**Heart** (cardiorespiratory failure).

**Total 20 marks**

## Test Five

A 70 year old woman is brought into hospital as an emergency admission. Her daughter believes that her mother fell and struck her head. An examination of the scalp indicates a laceration four centimetres in length overlying the right temple. The mother is unconscious.

1) If she has suffered a haemorrhage name the four spaces or compartments deep to the cranium into which she might have bled. (4 marks)

**Epidural space.**
**Subdural space.**
**Subarachnoid space.**
**Intraparenchymal area.**

2) During the history taking you learn from the daughter that the unconscious patient has a medical history of adult polycystic kidney disease. How is adult polycystic kidney disease inherited and which type of haemorrhage do you now favour? (3 marks)

**Autosomal dominant** (2 marks)
**Subarachnoid haemorrhage** (1 mark)

Adult polycystic kidney disease is associated with cysts in the liver, pancreas and lungs. The protein abnormality is polycystin which has a role in protein-protein interactions in the extracellular matrix. Hence it may have a role in blood vessel wall integrity. So it is not surprising that adult polycystic kidney disease predisposes to **berry aneurysms** in the **Circle of Willis**.

3) Name four clinical manifestations of this type of haemorrhage. (4 marks)

Any four of:
**Worst ever headache.**
**Altered consciousness.**
**Neck pain/stiffness.**
**Visual disturbances.**
**Speech changes (dysarthria).**
**Mood changes.**
**Movement disorders (ataxia).**
**Seizures.**
**Death.**

4) Name the two major arteries that directly supply the Circle of Willis. (2 marks)

Left and right **internal carotid** arteries
**Basilar** artery (from left and right **vertebral arteries**)

5) The daughter is upset by the state of her mother but is also anxious because she is eight months pregnant. If she delivers the baby at nine months, what are the definitions of the ranges of normal birth weights, low birth weights, very low birth weights and extremely low birth weights, into which her child might fall? (4 marks)

**>2500g is normal**
**<2500g is low**
**<1500g is very low**
**<1000g is extremely low**

6) Considering the foetal aetiology of adult disease, low birth weight has been directly linked to several *common* chronic adult illnesses/diseases. Name three. (3 marks)

Any three of:
**Insulin resistance and type 2 diabetes mellitus.**
**Coronary heart disease (ischaemic heart disease).**

**Hypertension.**
**Osteoporosis.**

**Total 20 marks**

## Test Six

A 78 year old male who has worked in heavy industry all his life presents with 5 kg of recent weight loss, chest pain that is worsened by breathing and stony dullness of the right lower lung field to percussion. His General Practitioner, GP, is concerned.

1) Which two diagnoses do you think that his GP is most concerned about? (2 marks)

**Mesothelioma.**
**Intrapulmonary lung malignancy (primary or secondary).**

> The GP is concerned about a potential malignancy – although the tumour appears to be affecting the pleural cavity, it may have started in the pleura or have started in the lung tissue *and extended into the pleura.* When considering cancer in any organ always consider the possibility of both ***primary*** and ***secondary*** malignancies. Students have a tendency to only think of primary malignancies.

2) What is the descriptive term for the chest pain he is experiencing? (1 mark)

**Pleuritic chest pain** is the pain that is directly associated with breathing movements.

3) A pleural aspirate is taken and sent to the cytopathology department. If the original lesion is pleural in origin, which cell type is likely to be the dominant presence? (1 mark)

**Mesothelial** cells.

> The pleural tissue consists predominantly of **mesothelial** cells so a malignancy of the pleura is a **mesothelioma**.

4) If the lesion is pleural in origin, state the most likely risk factor, and subdivide this into its two types. (3 marks)

**Asbestos fibres** – **chrysotile** and **amphiboles**.

> Asbestos fibres are the strongest risk factor for mesothelioma. Amphiboles have a greater durability and an association with iron that make them more **carcinogenic**.

5) Surprisingly, the cytopathology report indicates adenocarcinoma cells. Name four other different histological types of malignant lung neoplasms. (4 marks)

Any four of:
**Small cell carcinoma (= oat cell carcinoma)**
**Squamous cell carcinoma**
**Large cell/Neuroendocrine/Anaplastic carcinoma**
**Carcinoid tumour**
**Bronchioloalveolar carcinoma**
**Metastatic adenocarcinoma (e.g. colorectal, breast, renal *etc*).**
**Metastatic squamous cell carcinoma**
**Metastatic malignant melanoma**
**Metastatic lymphoma**
**Sarcoma**

> When answering questions about cancers students are advised to remember the following mantra "**Benign, malignant - primary or secondary**." Sometimes *cancer* is used to refer to carcinoma in situ that is in fact a *benign* lesion. Even if the word cancer is used correctly to refer to a malignancy, the list of possible answers can be significantly enlarged by remembering to consider secondaries; metastases to the organ in question. Always consider the common malignancies when generating lists of possible metastases. For example, lung, breast, prostate, cervical and skin cancers.

6) It is believed that the patient has a pleural effusion. Name four other fluids that can accumulate in the pleural space and indicate the descriptive term for each condition. (4 marks)

**Air – Pneumothorax** (air is a fluid!).
**Blood – Haemothorax.**
**Pus – Pyothorax or empyema.**
**Lymph – Chylothorax.**

7) On the side of the thorax with a pneumothorax, indicate how each of the following signs are likely to change (5 marks):

*Chest wall movement.*
*Percussion.*
*Breath sounds.*
*Vocal resonance.*
*Mediastinal shift.*

**Reduced chest wall movement.**
**Normal/hyper resonant percussion on the side of the pneumothorax.**
**Reduced breath sounds on the side of the pneumothorax.**
**Reduced vocal resonance on the side of the pneumothorax.**
**Mediastinal shift away from the side of the pneumothorax.**

**Total 20 marks**

## Test Seven

A longstanding type 1 diabetic presents with signs and symptoms of renal impairment. He is referred to a specialist nephrologist. The nephrologist takes a 24 hour urine sample and notes a total of 3.5g of protein.

1) What is the diagnosis and what is the most likely underlying cause? (2 marks)

**Nephrotic syndrome.**
**Diabetes mellitus.**

> The commonest secondary cause of nephrotic syndrome in adults is diabetes mellitus. Diabetes mellitus is the leading cause of end stage renal failure in Europe and Northern America.

2) What are the other major classic signs and symptoms associated with such a large protein loss in the urine? (3 marks)

Any three of:
**Oedema**
**Hypoalbuminaemia**
**Hyperlipidaemia**
**Sodium retention**
(The question is asking for the key features of nephrotic syndrome).

3) Name four other conditions/diseases that can cause such a large protein loss. (4 marks)

Causes of nephrotic syndrome - any four of:
**Minimal change glomerulonephropathy**
**Membranous glomerulonephritis**
**Membranoproliferative glomerulonephritis**
**Focal segmental glomerulosclerosis**
**Post infective glomerulonephritis**
**Diabetes Mellitus**
**Amyloid**
**Neoplasia – lymphoma/carcinoma**
**Endocarditis**
**Polyarteritis nodosa**
**Systemic lupus erythematosus**
**Sickle cell anaemia**
**Malaria**
**Drugs (penicillamine, gold)**
**IgA nephropathy**

> Nephrotic syndrome usually occurs as a result of structural changes in the glomerular basement membrane.

4) The biopsy of this diabetic's kidney shows abnormal malignant lymphocytes. There is a suspicion of lymphoma. What are the seven essential alterations for malignant transformation? (7 marks)

***Self-sufficiency in growth factors.***
***Insensitivity to inhibitory growth signals.***
***Evasion of apoptosis.***
***Defects in DNA repair.***
***Limitless replicative potential.***
***Sustained angiogenesis.***
***Ability to invade and metastasize.***

> This list is sometimes referred to as the "**magnificent seven**" changes required for malignancy. It is consistent with the multistep hypothesis for oncogenesis.

5) Which oncogene overexpression in breast cancer represents an overactive EGF receptor? (2 marks)

**c-erbB2** (or **Neu**) that codes for HER2 (Human Epidermal Growth Factor Receptor 2).

6) What is the role of the retinoblastoma protein in oncogenesis? (2 marks)

It is a **tumour suppressor gene** that normally binds to **E2F**. If a mutation to the retinoblastoma protein occurs, E2F is unbound and then becomes active in promoting transcription of products that activate the cell cycle.

**Total 20 marks**

## Test Eight

An 80 year old man is admitted to hospital with a history of peripheral vascular disease and hypertension. He appears to have developed new focal neurological signs.

1) "*Acute focal neurological deficit of vascular origin*" refers to which disease or disorder? (1 mark)

A **cerebrovascular accident (CVA)** or **stroke.**

2) Name three lifestyle risk factors for a stroke. (3 marks)

Any three of:
**Obesity/inactivity**
**Smoking**
**Alcohol**
**Contraceptive pill**
**Hormone Replacement Therapy**
**Syphilis**

3) Name five types of emboli other than thromboemboli. (5 marks)

Any five of:
**Cholesterol** – from atheroma.
**Fat emboli** – from fractures of long bones.
**Bone marrow** – from fractures of long bones.
**Air emboli** – from chest wall injury or obstetric procedures; 100 ml for a clinical effect.
**Nitrogen emboli** – from decompression sickness.
**Amniotic fluid emboli** – from a tear in placental membranes and rupture of uterine veins.
**Malignant cells –** which may occur as part of metastasis.
**Mycotic (infective)** – classically from a fragment of infective endocarditis.
**Foreign body** – bullets are a dramatic example!

4) *"Acute focal neurological deficit of vascular origin that lasts less than 24 hours"* refers to what disease or disorder? (1 mark)

**Transient ischaemic attack, TIA.**

TIA is a major risk factor for a subsequent stroke.

5) On taking a detailed history and examination, it becomes clear that the *new neurological signs* in the 80 year old man have disappeared. However, he demonstrates stable signs and symptoms that he has had since he was 50 years old. These include a festinant gait, micrographia and a tremor. Indicate which part of the brain has the primary lesion and which neurons are affected. (3 marks)

**Substantia nigra in the basal ganglia.** (2 marks)
**Dopaminergic (dopamine) neurons.** (1 mark)

6) What class of drug is entacapone and how is it able to treat the disease described in (5)? (3 marks)

It is a **COMT** (catechol-O-methyltransferase) **inhibitor** that can be used in combination with levodopa to **increase levels of dopamine** in the CNS by **inhibiting breakdown in the periphery**.

7) Why is bromocriptine useful in the management of this disease and to what class of drug does bromocriptine belong? (2 marks)

This synthetic dopamine agonist can replace dopaminergic neurone loss by acting at **dopamine receptors**. It is effective in younger patients and complements levodopa.
**Bromocriptine is a dopaminergic agonist at D2 receptors.**

8) Explain why anticholinergics are useful in the management of this disorder and name a drug that is used. (2 marks)

**Loss of dopaminergic neurons leads to a rise in CNS acetylcholine** which can itself cause symptoms. These additional symptoms can be blocked by anticholinergics:
**Trihexphenidyl (Benhexol, Broflex)**
**Benatropine**
**Orphenadrine**

**Total 20 marks**

## Test Nine

A 25 year old woman presents with jaundice to her GP.

1) What type of disease is usually responsible for pre-hepatic jaundice? (1 mark)

**Haemolysis/haemolytic disease.**

| Hereditary Haemolytic Anaemias | Acquired Haemolytic Anaemias |
|---|---|
| **Membrane**<br>Hereditary spherocytosis<br>Hereditary elliptocytosis<br>**Metabolic**<br>Glucose-6-phosphate dehydrogenase deficiency<br>Pyruvate kinase deficiency<br>**Haemoglobin abnormality**<br>Sickle cell anaemia<br>Thalassaemia | **Immune**<br>Autoimmune<br>Alloimmune - transfusion<br>Drug associated<br>**Red cell fragmentation syndromes**<br>**March haemoglobinuria**<br>**Infections**<br>Malaria, Clostridia<br>**Chemical/Physical agents**<br>**Secondary to renal or kidney disease**<br>**Paroxysmal nocturnal haemoglobinuria** |

2) Outline the faecal and urine colour changes of pre-hepatic jaundice. (2 marks)

**Dark(er) faeces.**
**Normal urine colour.**

| Colour | | | |
|---|---|---|---|
| **Faeces** | **Urine** | **Serum biochemistry** | **Interpretation** |
| Dark | Normal | Unconjugated Hyperbilirubinaemia | Usually due to haemolysis. |
| Pale | Dark | Conjugated Hyperbilirubinaemia | Cholestasis usually due to biliary obstruction. |
| Pale | Dark | Mixed Hyperbilirubinaemia | Often acute hepatitis |

3) Name four common signs of hepatic impairment. (4 marks)

Any four of:

**Oedema**
**Ascites**
**Haematemesis**
**Spider naevi/gynaecomastia**
**Purpura and bleeding**
**Coma**
**Infection**

4) Although the jaundice resolves, the cause of the jaundice is not identified. The GP is concerned that the underlying disease may progress to cirrhosis, through cycles of relapse and remission. What are the three major complications of cirrhosis? (3 marks)

**Liver failure, portal hypertension and liver cell carcinoma.**

Fibrosis predisposes to the development of malignancy.

5) This 25 year old woman is 35 weeks into her pregnancy. What is the normal undistressed foetal heart rate? (2 marks)

**110 – 160 bpm.**

6) Name the five classic ways in which presentation of a foetus occurs. (5 marks)

**Cephalic (vertex), breech, brow, shoulder and face.**

7) What are the two key hormones of lactation? (2 marks)

**Prolactin and oxytocin.**

8) Unfortunately the neonate is born suffering from a nerve palsy that affects the eyes. Which cranial nerve supplies the orbicularis oculis? (1 mark)

**The facial nerve.**

**Total 20 marks**

# Test Ten

A 48 year old man with a biopsy confirmed rectal adenocarcinoma undergoes major surgery (an anterior resection). Although the surgery is technically successful the patient has a myocardial infarction.

1) A myocardial infarction under general anaesthesia is not an uncommon event. Why does this occur? (1 mark)

The general anaesthesia causes a **drop in blood pressure** which can decrease coronary artery perfusion. In unfortunate patients this can lead to cardiac ischaemia and a myocardial infarction.

2) Which structural proteins are used as serum diagnostic markers of a myocardial infarction? (2 marks)

Troponin **I** and troponin **T**.

3) In a normal cardiac muscle cell under resting conditions to what do Troponin I and Troponin T bind respectively? What is the primary function of the troponin-tropomyosin complex? (3 marks)

Troponin T **binds to tropomyosin**.
Troponin I **binds to actin in thin filaments**.
The **troponin-tropomyosin** complex inhibits muscle contraction by preventing the binding of actin to myosin to form functional **actomyosin which would hydrolyze ATP** to cause muscle contraction.

4) Despite successful diagnosis and treatment the patient develops an arrhythmia. In a normal heart, starting at the sinoatrial node, describe the idealized pathway of electrical transmission leading to ventricular contraction. (4 marks)

SA node => **AV node** => **Bundle of His** => **Purkinje fibres** => **Left ventricle**

5) Name and briefly describe the four major classes of anti-arrhythmic drugs that affect the depolarization or repolarization phases of the action potential. (4 marks)

**Vaughan-Williams Classification:**

- **Type I: blockage of voltage gated $Na^+$ channels**
- **Type II: β-adrenergic blockers**

- **Type III: action potential prolonging ($K^+$ channel)**
- **Type IV: calcium channel blockers**

6) Which ion channel is responsible for the *absolute refractory period* of cardiac myocytes, preventing the initiation of another action potential? (1 mark)

**$Na^+$ ion channel**

7) The arrhythmia that the patient develops is atrial fibrillation. What pulse characteristics are typical of atrial fibrillation? (1 mark)

**An irregularly irregular rhythm.**

8) Atrial fibrillation has implications for Virchow's triad. Describe Virchow's triad. (3 marks)

**Altered coagulation.**
**Altered blood flow.**
**Altered vessel walls.**

9) Name two physiological growth factors that can induce angiogenesis. (1 mark)

**Vascular Endothelial Growth Factor (VEGF)**
**Basic Fibroblast Growth Factor (bFGF)**
**Angiopoietin 1**
**Angiopoietin 2**

**Total 20 marks**

## Test Eleven

You are a junior doctor working in an oncology department. Your responsibilities include breaking bad news to patients and interpreting and communicating pathology reports.

1) A pathology report on a breast biopsy indicates a local breast cancer. What feature must the histopathologist have seen under the microscope ***to be sure*** that the tumour is a malignancy? (1 mark)

**Invasion** (at least through the basement membrane).
The definition of cancer requires the presence of invasion.

2) A previous CT scan shows that the affected patient has widely disseminated breast cancer. What are the eight essential steps that a malignant cell must undergo to successfully metastasize? (8 marks)

- **Invasion** of the basement membrane.
- Passage through the **extracellular matrix**.
- **Intravasation**.
- Avoidance of **immune interaction**.
- **Platelet** adhesion.
- **Adhesion to endothelium**.
- **Extravasation**.
- **Angiogenesis**.

3) What is another word with equivalent meaning to *metastases*? (1 mark)

**Secondaries.**

4) Complete the following table (3 marks):

| Benign | Malignant |
|---|---|
| Osteoma | **Osteosarcoma** |
| **Adenoma** | Adenocarcinoma |
| Melanocytic naevus | **Malignant melanoma** |

5) Classically what are the four common malignancies that have a predominantly osteolytic effect on bone? (4 marks)

**Breast cancer**
**Lung cancer**
**Thyroid cancer**
**Renal cancer**

The list is remembered as the **four Bs**: **B**ronchial, **B**reast **B**yroid and **B**idney. ☺

6) The patient is believed to carry the aetiological BRCA1 mutation. What class of neoplastic gene is represented by BRCA1? (1 mark)

**Tumour suppressor gene.** In this case it is a mutated and hence dysfunctional tumour suppressor gene.

7) The retinoblastoma gene is in this same functional class and unfortunately the patient is also carrying a mutation in this gene. Under physiological circumstances what is the major transcription factor to which the retinoblastoma protein binds to carry out its function? (1 mark)

**E2F**

> Unrestrained E2F acts to facilitate growth by triggering the synthesis of key proteins at a transcriptional level. Under physiological conditions retinoblastoma protein binds to E2F to inhibit its action. In effect the normal action of the retinoblastoma protein in binding to E2F is to prevent an excessive growth signal. Hence the retinoblastoma gene is considered a **tumour suppressor gene**.

8) Eventually the breast cancer progresses to the brain. As this brain cancer is inoperable only supportive management is possible. Dexamethasone is part of the management. To what class of drug does this belong and what syndrome may occur with chronic use? (1 mark)

**Glucocorticoid** (Corticosteroid) – 0.5 mark
**Cushing's syndrome** – 0.5 mark

**Total 20 marks**

## Test Twelve

A 50 year old man with longstanding rheumatoid arthritis visits his GP complaining of shortness of breath on moderate exertion. He appears cyanotic. Auscultation reveals basal crepitations without a wheeze. No stony dullness is found. No pyrexia is noted. The patient does not remember when the shortness of breath started but thinks it has been going on "for quite a while."

1) What is the characteristic histological feature of idiopathic pulmonary fibrosis (cryptogenic fibrosing alveolitis)? (1 mark)

**Fibroblastic focus (foci).**

2) Considering its effect on lung function, what class of lung disease is idiopathic pulmonary fibrosis? (1 mark)

**Restrictive lung disease.**

3) What are the classic complications of idiopathic pulmonary fibrosis? (4 marks)

Any four of:
**Infection**
**Emphysema**
**Cor pulmonale**
**Secondary pulmonary hypertension**
**Primary lung cancer**
**Respiratory failure**
**Death**

4) What type of respiratory failure is characterized by hypoxia and hypercapnia? Name two common causes of this type of respiratory failure. (3 marks)

**Type 2 respiratory failure**.
This type of respiratory failure is identified by hypercapnia often caused by obstructive lung disease and so patients may exhibit "blue bloater" clinical features. Classic causes are **COPD (emphysema, chronic bronchitis)**, severe obesity and malignancy.

5) What is the therapy for stage 2 (moderate) COPD? (2 marks)

**Tiotropium 18micrograms daily.**
**Symbicort 400/12 or seretide 500 inhaler twice daily.**
**Salbutamol or terbutaline inhaler (2-4 puffs) up to 6 times per day.**
**Rehabilitation.**

6) Name four other common causes of right heart failure. (4 marks)

Any four of:

Left heart failure.
Chronic lung disease.
Pulmonary embolism.
Valvular disease – usually mitral valve disease.
Right ventricular infarction.
Congenital heart disease (e.g. ASD or VSD).
Adult or acute respiratory distress syndrome.
Increased afterload complicating thoracic surgery.
Protamine administration.

7) One of the pharmacological treatments for heart failure is diuretic administration. What is the pharmacodynamic target of spironolactone and what is the particular clinical significance of this drug? (2 mark)

**The aldosterone receptor is the target of spironolactone. This a potassium sparing diuretic.**

8) Right heart failure is a cause of oedema. What type of oedema occurs? Name two other major organ impairments that can also cause oedema. (3 marks)

**Transudate.**
**Liver and kidneys.**

**Total 20 marks**

## Test Thirteen

A 40 year old daughter takes her 70 year old father to their GP, complaining that he has recently become more forgetful.

1) What is dementia? (2 marks)

A progressive **loss of higher cortical functions** (e.g. memory or reasoning) **without decreased consciousness**.

> Many diseases cause systemic effects that have a secondary effect of lowering consciousness and an associated decline in cognitive function. Because cognitive function returns as normal consciousness returns, such diseases are not regarded as dementias.

2) Name the two commonest disease causes of dementia. (2 marks)

**Alzheimer's disease.**
**Multi-infarct dementia (vascular dementia).**

3) What is the name of the validated, brief and commonly used clinical test that assesses dementia? What is the maximum possible score on this test? What is the range of scores that is defined as normal? (3 marks)

**Mini-mental state examination.**
**30 is the maximum score. 27-30 is the normal range.**

The mini mental state examination (MMSE) is used to assess cognitive impairment. It is most commonly used as a screening tool for dementia and to assess the progression of the dementia. A typical patient, without treatment, may have a declining score of approximately 3-4 points per year.

4) What are the characteristic CNS histological features of Alzheimer's disease? (3 marks)

**Neurofibrillary tangles**
**Amyloid plaques**
**Cortical atrophy**

5) Dementia is a chronic morbidity that is often associated with depression. What are the two commonly used screening questions for depression? (2 marks)

- During the past month have you often been bothered by feeling down, depressed or hopeless? (*low mood*)
- During the past month have you often been bothered by little interest or pleasure in doing things? (*anhedonia*)

6) Outline six common clinical manifestations of depression that may facilitate diagnosis. (6 marks)

Any six of:

(1) **Depressed mood** most of the day, nearly every day, as indicated by either subjective report (e.g. feels sad or empty) or observation made by others (e.g. appears tearful).

(2) Markedly **diminished interest or pleasure** in most activities nearly every day (as indicated by either subjective account or observation made by others).

(3) Significant **weight loss** when not dieting or weight gain (e.g. a change of more than 5% of body weight in a month), or decrease or increase in appetite nearly every day.

(4) **Insomnia** or hypersomnia nearly every day.

(5) **Psychomotor retardation or agitation** most days (observable by others, not merely subjective feelings of restlessness or being slowed down).

(6) **Fatigue** or loss of energy most days.

(7) Feelings of **worthlessness or excessive or inappropriate guilt** (which may be

delusional) most days.

(8) Diminished ability to think or **concentrate, or indecisiveness**, most days (either by subjective account or as observed by others).

(9) Recurrent **thoughts of death**, recurrent suicidal ideation without a specific plan, or a suicide attempt or a specific plan for committing suicide.

7) Name two commonly used psychological interventions in the management of depression. (2 marks)

Any two of:
Cognitive Behavioural Therapy.
Psychodynamic Therapy.
Individual guided self-help.
Group-based peer support.
Interpersonal therapy.
Behavioural activation.
Couples therapy.

**Total 20 marks**

## Test Fourteen

1) Which major nerve lies deep to the bicipital aponeurosis in the antecubital fossa? (1 mark)

The commonest path of the **median nerve** travels along the forearm deep to the bicipital aponeurosis in the antecubital fossa.

2) From which roots in the brachial plexus does the median nerve originate? (2 marks)

**C5, C6, C7, C8 and T1.**

3) What is the name of the branch of the median nerve that provides the motor supply to the thenar muscles? (1 mark)

The **recurrent branch** of the median nerve. This supplies the muscles of the thenar eminence – specifically the abductor pollicis brevis, flexor pollicis brevis and opponens pollicis.

4) Name four common causes of carpal tunnel syndrome. (4 marks)

Any four of:

| | |
|---|---|
| Anatomical factors | **Wrist fractures**<br>**Wrist dislocations** |
| Neuropathic comorbidity | **Diabetes mellitus**<br>**Alcohol abuse** |
| Inflammation (autoimmune or infective) | **Rheumatoid arthritis** |
| Fluid retention (systemic) | **Pregnancy**<br>**Menopause**<br>**Obesity**<br>**Hypothyroidism**<br>**Renal failure** |
| Repetitive movements of the wrist | **Keyboard work**<br>**Vibrating tools** |

(1 mark each for any four of the above).

> The carpal tunnel is a space that is limited in size by the flexor retinaculum and the carpal bones of the wrist. Hence fluid retention will increase pressure on the median nerve as it passes through the tunnel. Similarly inflammatory oedema (acute inflammation, chronic inflammation, repetitive strain, autoimmune disease) can also cause carpal tunnel syndrome.

5) Name four signs or symptoms that are commonly presented by a sufferer of carpal tunnel syndrome. (4 marks)

Any four of:

**Paraesthesia** (local tingling). **Local pain that may be worse at night.** **Muscle atrophy** (especially of thenar eminence). **Weak pinch grip.** **Inadvertent dropping of objects. Sensory loss** (that may be permanent).

6) Name three tests used to diagnose carpal tunnel syndrome. (3 marks)

Any three of:

**Tinel's sign. Phalen's test/maneuver. Reverse Phalen's test. Nerve conduction studies. Electromyograms.**

> A positive Tinel's sign causes distal tingling in the distribution of the median nerve as the result of light percussion over the carpal tunnel.

7) Describe the mechanism of action of hydrocortisone as an immunosuppressive and anti-inflammatory drug. (5 marks)

Any two of:
Hydrocortisone is a corticosteroid used for its anti-inflammatory and immunosuppressive effects. The mechanism of action involves binding to intracellular **steroid receptors that then migrate to the DNA to alter transcription**. One effect is to diminish the activity of phospholipase A2 – diminishing the synthesis of arachidonic acid and hence inflammatory mediators.

Hydrocortisone's anti-inflammatory action is believed to be due in part to the **suppression of migration of neutrophils** and reversal of the **increased capillary permeability of acute inflammation**.

Hydrocortisone's immunosuppressive effects are believed to be due to inhibition of **T cell proliferation** and **interleukin-2 production**.

**Total 20 marks**

## Test Fifteen

A 75 year old man presents to his GP complaining of pain on micturition. He has a history of recent urinary tract infections.

1) Why is this man likely to be predisposed to urinary tract infections? (2 marks)

At his age this man is likely to have an enlarged prostate (**benign prostatic hyperplasia/hypertrophy**) that would predispose to **urinary stasis and infection**.

2) What is the most likely source of the infecting bacteria and what type of bacteria are they likely to be? Name the most likely bacterium. (3 marks)

**Gastrointestinal tract** is the likely source of the bacteria.
**Gram negative bacilli** are the most likely type.
**Escherichia coli** is the most likely bacterium.

3) Unfortunately a prostatic biopsy indicates that the patient has a small peripheral focus of malignant tumour. Name the most probable histological type of malignancy. (2 marks)

**Primary**/prostatic **adenocarcinoma**.

4) What is the commonest grading system used for this malignancy? (1 mark)

**Gleason** grading

5) The patient complains of chronic loin pain, fever and lethargy. The GP notices that the patient is now hypertensive. What is the most probable diagnosis and most worrying complication? (2 marks)

**Chronic pyelonephritis**. (1 mark)
**Renal failure** or **septic shock** or **perinephric abscesses** can be worrying complications. (1 mark for any of the three)

6) Name three commonly used antibiotics in the treatment of urinary tract infections. (3 marks)

Any three of:
**Trimethoprim**
**Amoxicillin**
**Nitrofurantoin**
**Ciprofloxacin**
**Levofloxacin**

7) Name four other species of bacteria that commonly cause UTIs and indicate the commonest infecting scenario for each (4 marks):

Any four bacteria:

| Bacteria | Common scenario for infection |
|---|---|
| *Escherichia coli* | *Originating from gastrointestinal tract – community or hospital acquired.* |
| Klebsiella pneumonia | **Hospital acquired** and associated with indwelling catheters |
| Proteus mirabilis | **Hospital acquired** and older patients |
| Pseudomonas aeruginosa | **Hospital acquired** |
| Staphylococcus saprophyticus | Associated with **sexual intercourse** |
| Enterobacter aerogenes/cloacae | **Hospital acquired** and immunosuppressed |

8) Describe the mechanism of action of trimethoprim. (3 marks)

It inhibits **dihydrofolate reductase** in bacteria, inhibiting **folate metabolism**, preventing nucleotide synthesis and hence **inhibiting DNA synthesis**.

**Total 20 marks**

# Test Sixteen

A 37 year old man is an emergency admission to hospital with suspected meningitis.

1) Name the three layers of the meninges. (3 marks)

**Pia mater**
**Dura mater**
**Arachnoid mater**

2) Assuming that this is a bacterial meningitis, which two bacteria are the most probable aetiological agents? (2 marks)

**Neisseria meningitidis** and **Streptococcus pneumoniae** are the most common causes in adults (Haemophilus influenza and Group B streptococci are more common in children and neonates respectively).

Although the patient has some meningeal signs he starts to display odd behaviour and a lowered level of consciousness. You believe that he may be suffering from encephalitis.
3) What type of encephalitis would be of greatest concern in the acute setting and why? (2 marks)

**Herpes Simplex** Encephalitis.
This is a **medical emergency** requiring urgent administration of acyclovir (aciclovir).

4) What is the mechanism of action of the leading drug used in the treatment of this type of encephalitis? (2 marks)

Aciclovir is a **nucleoside analogue** which ultimately inhibits the **viral DNA polymerase**.

5) Complete the following table (8 marks):

| | Normal CSF | Bacterial Infection | Viral Infection |
|---|---|---|---|
| Cells | 0-5 wbc/ml | >1000 wbc/ml | <1000 wbc/ml |
| Polymorphs | 0 | **Predominate** | **A few early** |
| Lymphocytes | 5 | **Only appear late** | **Predominate** |
| Glucose | 40-80mg/dl | **Decreased** | **Normal** |
| (CSF:plasma) glucose ratio | 66% | **<40%** | Normal |
| Protein | 5-40 mg/dl | **Increased** | Negligible change |
| Culture | Negative | Positive | Negative |
| Gram staining | Negative | Positive | Negative |

6) If the treatment of the encephalitis (or meningitis) is too late or insufficient, raised intracranial pressure may occur. This can lead to fatal herniation. One of the subtypes of herniation is transtentorial (central). Name the structure over which this herniation occurs and indicate the layer of the meninges from which it is derived. (3 marks)

The **tentorium cerebelli** is derived from the **dura** mater and separates the occipital lobes from the cerebellum.

**Total 20 marks**

## Test Seventeen

Alcohol abuse has a range of deleterious short term and long term effects. As a result, assessing alcohol abusers is an important clinical task.

1) What four screening questions are commonly used to identify alcohol abusers? (4 marks)

**C** Have you felt the need to **Cut** down on your drinking?
**A** Have you been **Annoyed** by criticism of your drinking?
**G** Have you felt **Guilty** about your drinking?
**E** Have you taken an **Eye-opener** in the morning?

2) What are the recommended safe maximum number of units of alcohol that are permitted for men and women in daily and weekly terms? (4 marks)

**Women**
Not more **than 2-3 units per day** (previously **< 15 units per week**).
**Men**
Not more than **3-4 units per day** (< **20 units per week**).

3) Name five gastrointestinal malignancies at a greater risk of occurring in alcohol abusers. (5 marks)

**Any five of**:
Oral cancer (squamous cell carcinoma)
Pharyngeal cancer (squamous cell carcinoma)
Laryngeal cancer (squamous cell carcinoma)
Oesophageal cancer (squamous cell carcinoma)
Gastric cancer (adenocarcinoma)
Liver cancer (hepatocellular carcinoma)
Colorectal cancer (adenocarcinoma)

As a junior doctor on an oncology ward you have to manage terminally ill patients.

4) Define the doctrine of double effect as it applies to clinical practice. (3 marks)

The *doctrine of double effect* states that it is morally permissible to **perform an action with a good effect but also a bad effect if the prescribed conditions are satisfied**. The classic clinical example is the **administration of morphine** to a terminally ill patient with a painful malignancy – the morphine will relieve the pain as its primary purpose, however, there is also the risk of **respiratory depression** and death of the patient.

5) Classically under what moral conditions is the doctrine of double effect applicable? (4 marks)

- The action itself is morally good, or at least morally neutral.
- The agent's intention is good.
- The good effect does not follow from the bad effect.
- There is a proportionately grave reason for the action in question.

**Total 20 marks**

## Test Eighteen

1) Define disease incidence. (1 mark)

**Number of new cases of a disease arising in a defined population in a defined period of time.**

2) When assessing the health of a population, it is often necessary to rely on surveys of patients. What are the two biggest issues regarding the accuracy of health information acquired in this way? (2 marks)

**Responder bias (only certain types of patient respond).**
**Reporting bias (patients give subjective responses).**

3) Define attributable risk. (2 marks)

**Attributable risk = (incidence of disease in exposed group - incidence of disease in unexposed group) ÷ incidence of disease in exposed group**

4) List the Bradford Hill's criteria for inferring causality (9 marks).

|  |  |
|---|---|
| Strength of association | **Association features** |
| Specificity of association | |
| Consistency of association | |
| Temporal sequence | **Exposure / outcome** |
| Dose response | |
| Reversibility | |
| Coherence of theory | **Other evidence** |
| Biological plausibility | |
| Analogy | |

5) Explain what is meant by the *general fertility rate*. (2 marks)

**Number of live births per 1000 women, aged between 15 and 44 years (child-bearing years).**

6) Explain what is meant by the phrase "population census." (2 marks)

**The simultaneous recording of demographic data by the government at a particular time, pertaining to all the persons living in a particular territory.**

**7)** What is the WHO definition of health? (2 marks)

**"A state of complete physical, social and mental wellbeing and not merely the absence of disease or infirmity."**

**Total 20 marks**

## Test Nineteen

A teenager presents at his GP surgery complaining of new onset shortness of breath on exertion with an obvious wheeze.

1) Name three common atopic disorders. (3 marks)

Any three of:
**Asthma**
**Hayfever**
**Eczema**
**Anaphylaxis**

2) List four histological features of asthma. (4 marks)

Any four of:
**Smooth muscle hypertrophy of the bronchial tree.**
**Mucous gland hyperplasia/hypertrophy**
**Airway oedema**
**Inflammatory cell infiltrate**
**Epithelial damage/basement membrane thickening**

3) What functional class of lung disease is asthma? (1 mark)

**Obstructive lung disease.**

4) The young boy has just started smoking. If he continues to smoke what non-neoplastic lung diseases is he prone to? (3 marks)

Any three of:
Chronic Obstructive Pulmonary Disease – **Emphysema** and **Chronic Bronchitis**
**Respiratory bronchiolitis-interstitial lung disease (RB-ILD)**
**Desquamative interstitial pneumonia**
**Langerhans' cell histiocytosis**

5) Which of these diseases (question 4) is most likely to be exacerbated by alpha-1 antitrypsin deficiency? (1 mark)

**Emphysema.**

6) Chronically what effect will alpha-1-antitrypsin deficiency ultimately have on the liver? (2 marks)

**Cirrhosis** and **liver failure**.

> Alpha-1-antitrypsin inhibits protease activity and so acts as a brake on inflammatory proteolysis. A deficiency of alpha-1 antitrypsin will permit more extensive tissue damage during any inflammatory process. Hence the liver will progress to a cirrhotic state far sooner in the face of such a deficiency.

7) What colour changes are likely to occur to the faeces and urine of such an individual? (2 marks)

**Pale faeces, dark urine.** Classically, dark faeces and normally coloured urine suggest a haemolytic cause of hyperbilirubinaemia (this is a pre-hepatic cause). In contrast, pale faeces and dark urine imply an obstructive cause of post-hepatic jaundice. Clinically it is difficult to distinguish the intrahepatic causes of raised bilirubin from the obstructive causes – based on the features of urine and faeces alone:

| **Colour** | | | |
|---|---|---|---|
| **Faeces** | **Urine** | **Serum biochemistry** | **Interpretation** |
| Dark | Normal | Unconjugated hyperbilirubinaemia | Usually due to *haemolysis*. |
| Pale | Dark | Conjugated hyperbilirubinaemia | Cholestasis usually due to biliary *obstruction*. |
| Pale | Dark | Mixed hyperbilirubinaemia | Classically acute hepatitis (*intrahepatic* pathology). |

8) Haemochromatosis and Wilson's disease can ultimately cause similar changes in the liver. What are the underlying disease processes in each case? (4 marks)

Haemochromatosis occurs as a result of **excessive accumulation of iron** in the human body – it may be primary (endogenous) or secondary (e.g. as a result of excessive blood transfusion). The iron deposition predisposes to fibrosis and hence **liver cirrhosis**.

Wilson's disease occurs a result of **copper deposition in the liver because of defective excretion of copper**. This can result in fatty change, acute hepatitis, chronic hepatitis, **cirrhosis** or hepatic necrosis.

**Total 20 marks**

## Test Twenty

A 47 year old amateur sportsman was admitted to his local Accident and Emergency department after complaining of chest pain. It is Sunday night and he spent the afternoon playing squash. The previous day he played football for 3 hours. The chest pain is described as 7/10 in intensity, central and radiates to his neck. It started two hours after he finished his game of squash, and is unusual in not being relieved by ibuprofen. The pain has continued and is still present on admission.

He has no family or personal history of arrhythmia and no evidence of hypertrophic obstructive cardiomyopathy (HOCUM or HOCM) and no evidence of hyperlipidaemia. He has never smoked. He has no personal or family history of diabetes mellitus. His parents are in their 70s and are in good health – neither has suffered from angina or a myocardial infarction.

Palpation on the ribs or sternum could not elicit the pain.

The junior doctor managing the patient decides that the most likely diagnosis is a myocardial infarction.

1) Name four drugs routinely used in the immediate management of a myocardial infarction. (4 marks)

**Oxygen, Nitrates, Clopidogrel, Aspirin, Morphine, Enoxaparin (or equivalent drug), Streptokinase and Tissue Plasminogen Activator.**

**R**eassurance
**O**xygen
**M**orphine with an antiemetic drug.
**A**spirin
**N**itrate (usually glyceryl trinitrate).
**C**lopidogrel (or Ticagrelor)
**E**noxaparin (or equivalent drug e.g. fondaparinux).

2) Clopidogrel and enoxaparin share a major side-effect (adverse effect). Name this adverse effect. (1 mark)

An increased tendency to **haemorrhage**.

3) Describe the mechanism of action of enoxaparin. (3 marks)

Enoxaparin is a **low molecular weight heparin** that can be administered in a subcutaneous manner. Heparin's primary mechanism of action requires **binding to antithrombin III to increase its activity**. Heparin causes a conformational change in antithrombin III that increases its **binding to the active forms of key clotting factors i.e. 12a, 11a, 10a, 9a and 7a**. Antithrombin III inhibits their actions by binding irreversibly to them and so prevents the action of the clotting cascade.

The junior doctor requests an ECG and takes blood to test for the release of cardiac enzymes/troponin. The ECG changes are indeterminant with no clear ST elevation. However, troponin and CK-MB levels are significantly raised in the venous blood.

4) Why would a myocardial infarction raise circulating levels of troponin and CK-MB? (1 mark)

**Death of cardiac myocytes** results in the breakdown of the cell membrane and the release of intracellular contents into the adjacent tissue, that can then enter the systemic circulation.

5) Although the patient finds the drug regime effective, he complains that one of the medications is giving him a headache. Which medicine is it likely to be and why is the headache occurring? (2 marks)

**Glyceryl trinitrate**, GTN. (1 mark)
Headaches are believed to occur because of the GTN mediated elevation of **nitric oxide** concentration resulting in smooth **muscle relaxation and vasodilation**. (0.5 mark each)

6) How and why are streptokinase and tissue plasminogen activator, tPA, useful in the management of a myocardial infarction? (4 marks)

They are both **thrombolytic drugs** that breakdown the acutely occluding **blood clot** in the coronary artery.

Streptokinase – **facilitates the conversion of plasminogen to plasmin that lyses fibrin in blood clots.**
tPA – **catalyzes the conversion of plasminogen to plasmin.**

Three days after admission the patient is still suffering from a similar pain and has developed a pyrexia of 38°C. Angiography indicates that the three major coronary arteries are each more than 90% patent. The relatively young age and paucity of risk factors caused the consultant supervising the junior doctor to reconsider the working diagnosis.
7) What is your favoured new diagnosis? (2 marks)

**Myocarditis** is the best diagnosis. (2 marks)
**Pericarditis** is less likely. (0.5 mark)
(Myocarditis and pericarditis can occur together).

8) What medications would you use to manage the latest diagnosis? (3 marks)

Reasonable combinations from:
**1) None** – most patients recover without drug intervention. (1 mark) *or*
**2) Paracetamol and NSAID** – antipyrexial and analgesic. (2 marks) *or*
**3) Paracetamol/NSAID with digoxin, frusemide, dobutamine or ACE inhibitor** – heart failure medications. (3 marks)

Often the cause of the myocarditis is not found and most patients are managed supportively. Generally, physicians are reluctant to take a cardiac biopsy from a patient suffering cardiac chest pain – although this could lead to a definitive diagnosis. The causes, when identified, are usually autoimmune or infective:

Viral (e.g. Coxsackie virus, HIV, rubella virus, cytomegalovirus)
Protozoan (e.g.Trypanosoma cruzi causing Chagas disease)
Bacterial (e.g. Brucella)
Fungal (e.g. Aspergillus)
Parasitic (e.g. Schistosoma)

**Total 20 marks**

## Test Twenty-One

It is 11.30pm on a Monday night in a delivery suite at the local teaching hospital. A husband is holding his wife's hand as she enters the 22nd hour of labour. Despite the mother's small to medium build, the child appears large and this seems to be making the delivery difficult. The midwife has a student nurse assisting in the suite. Being an exceptionally well trained and gifted student she remembers details of her *introduction to biochemistry* course.

1) On the course she remembers being told a word for *big babies*. What is the word most likely to have been? (1 mark)

**Macrosomia** (1 mark)

2) The nursing student remembers that there was a common endocrinological cause for such large babies – gestational hyperglycaemia. Name four hyperglycaemic hormones. (4 marks)

Any four of:
**Glucagon.**
**Growth hormone.**
**Adrenaline.**
**Noradrenaline.**
**Lactogen.**
**Cortisol.**
**Thyroxine.**
**Triiodothyronine.**
**Oestradiol (any oestrogen).**
**Progesterone (any progestogen).**
(1 mark each)

3) Accordingly, to what disorder is such a pregnant woman prone? (1 mark)

**Diabetes mellitus** (1 mark)

During the long labour the mother-to-be has plenty of time to think. Her mind drifts to her 22 year old male cousin. She was close to her cousin and had been quite distressed when six weeks ago he lost a lot of weight very quickly, started to drink a lot of fluids and was repeatedly going to the toilet. She noticed that her cousin's breath also smelt funny and his eyes looked sunken. Eventually, after being persuaded to visit his GP, he was rushed into hospital

4) What is the name for the condition that the young man was in and which protein was part of the core treatment? (2 marks)

**Diabetic Ketoacidosis (DKA)** (1 mark)
**Insulin** (1 mark)

5) Why had he lost weight and developed sunken eyes? (2 marks)

The weight loss and sunken eyes were the result of **dehydration** which itself was caused by a severe **osmotic diuresis as a result of the hyperglycaemia.** (1 mark each for phrases similar to the highlighted areas. 0.5 mark for any reference to fat loss or muscle loss.)

6) Name the cause of the smell and give two examples of substances that could have been responsible for the smell. (3 marks)

The smell was caused by **ketone bodies**. (1 mark)

Examples of ketone bodies are **acetoacetate** and **β-hydroxybutyrate** (neither of these substances are actually ketones so "ketones" cannot be accepted as an answer to the first part of this question). The third possible answer is **acetone**, which actually *is* a ketone. (2 marks for any two, 1 mark each)

7) What disease is her cousin likely to be manifesting? (2 marks)

***Type 1* diabetes mellitus** (the question awards two marks so *diabetes mellitus* alone is an insufficient answer). One mark for Type 1 and one mark for diabetes mellitus.

8) i) Is 12 mM a normal physiological concentration of blood glucose?
ii) Is 4.2 mM a normal physiological concentration of blood glucose?

iii) What initial effect does insulin have on the plasma potassium concentration?
iv) How does insulin have this effect on plasma potassium concentration?
(5 marks)

i) **No.** A random blood glucose of **12 mM** is diagnostic for diabetes mellitus, so this is a pathological concentration. (1 mark)
ii) **Yes.** Normal blood glucose has an **average value of about 5 mM** (90 mg/dL) and a value below 3.5 mM often causes symptomatic hypoglycaemia. (1 mark)
iii) Insulin **decreases the plasma potassium** concentration. (1 mark)
iv) Insulin stimulates the **sodium potassium pump (Na/K-ATPase)** in the cell membrane which acts as an **antiporter, exporting sodium ions into the extracellular fluid and importing potassium ions** into the intracellular fluid. (2 marks)

**Total 20 marks**

## Test Twenty-Two

It is 8pm and you are a junior doctor called to see a 33 year old female patient who has a bad headache with associated nausea, vomiting and photophobia. She has been an inpatient for two weeks for the management of recurrent ulcerative colitis. She had been able to eat a lunch that included chocolate cake and lemon sorbet. The headache has continued for 8 hours despite her nurse giving her two 500mg paracetamol (acetaminophen) tablets. Her blood pressure and pulse are normal. On examination you are unable to elicit signs of meningism and no fever is noted.

1) Based on this information what is the most likely diagnosis? (1 mark)

**Migraine headache** (1 mark)

2) The most likely diagnosis can classically be divided into four phases. Name the four phases. (4 marks)

| | | |
|---|---|---|
| **Premonition** | | **Prodrome** |
| **Prodrome** | or | **Aura** |
| **Headache** | | **Headache** |
| **Postdrome** | | **Postdrome** |

(1 mark each)

Not all migraine headaches have a discernible prodrome. So the patient does not always know the headache is coming. Migraine without aura usually consists of a severe headache greater than four hours of duration and is associated with nausea/vomiting or photophobia/phonophobia. Classic triggers of migraine include *c*itrus fruit, *c*affeine, *c*hocolate, *c*heese, the *c*ontraceptive pill and al*c*ohol ($\mathbf{C}_2H_5OH$).

3) In general terms what pathophysiological process in the brain is directly associated with the aura? (2 marks)

**Cortical depression.** (2 marks) **or**
**Artery spasm.** (2 marks)

4) Name three common different types of headache. (3 marks)

Any three of:
**Tension**
**Cluster**
**Migraine**
**Sinus/sinusitis**

5) Which part of the brain is believed to be the migraine generator? (1 mark)

**Brainstem** (1 mark) or
**Midbrain** (1 mark) or
**Dorsal raphe, locus ceruleus and periaqueductal gray matter** (1 mark)

6) It is believed that the pathophysiology of this severe headache involves inflammatory neuropeptides. Based upon this information what class of drug would you use to treat this headache? (2 marks)

**NSAIDS, Non-steroidal anti-inflammatory drugs.** For example, ketorolac.

7) The most likely diagnosis of the headache is also associated with epilepsy. Name four established causes of epilepsy. (4 marks)

Any four of:
**Developmental abnormalities (e.g. cerebral palsy)**
**Trauma to the skull**

**Stroke (brain infarcts)**
**Brain tumours**
**Encephalitis/meningitis**
**Alcohol abuse**

However, most cases of epilepsy are idiopathic.

8) The unfortunate patient goes on to develop epilepsy that is recurrent and resistant to treatment. Seizures occur in the daytime and affect the level of consciousness. (a) What ethical principle governs your reporting this patient to the governmental authorities responsible for road safety? (b) What duty must you contravene under such circumstances? (2 marks)

**(a) Doctors have an overarching duty of care to protect the public. This is essentially a principle of justice that requires the doctor to care for the majority.** (1 mark)
**(b) The doctor is obliged to breach their duty to protect the patient's confidentiality.** (1 mark)

9) If doctors report their patients' misdeeds to the authorities they may cease to be trusted by their patients. Patients will be less willing to seek out doctors and trust them – damaging the doctor-patient relationship. If the doctor bases his decision not to report the epileptic patient on the likely subsequent effects, what ethical morality/view is being demonstrated? (1 mark)

**Consequentialism.** (1 mark)

**Total 20 marks**

## Test Twenty-Three

An enthusiastic weekend cyclist attends visits his general practitioner to complain about wrist pain and numbness in the lateral two digits of the affected hand whilst he is cycling. He uses drop-down handlebars and usually cycles for at least 90 minutes continuously during each session. This 37 year old man is otherwise fit and well with no significant medical history.

1) Which cords give rise to the median nerve? (2 marks)

**Lateral cord** (lateral root).
**Medial cord** (medial root).

2) Name the five terminal branches of the brachial plexus. (2.5 marks)

**Radial, median, ulnar, musculocutaneous and axillary nerves.** (0.5 mark for branch)

3) Name the intrinsic muscles of the hand supplied by the median nerve. (2.5 marks)

**L** Lumbrical 1, Lumbrical 2
**O** Opponens pollicis
**A** Abductor pollicis brevis
**F** Flexor pollicis brevis (0.5 mark each).

4) Why do you think that the cyclist is experiencing this pain? Briefly explain the pathogenesis. (2 marks)

The cyclist is likely to be putting **direct pressure on the carpal tunnel** because of the way he is **holding the handle bars** – indirectly affecting the median nerve. When he stops holding the bars, the pressure on the median nerve is relieved and the symptoms abate.

5) The cyclist has positive results to Tinel's test, Phalen's test and Reverse Phalen's test. From which syndrome is the cyclist likely to be suffering? (1 mark)

**Carpal Tunnel Syndrome**

6) Name four classic causes of this syndrome. (4 marks)

Carpal Tunnel Syndrome can classically be caused by any of:

| | |
|---|---|
| Anatomical factors | **Wrist fractures**<br>**Wrist dislocations** |
| Neuropathic comorbidity | **Diabetes mellitus**<br>**Alcohol abuse** |
| Inflammation (autoimmune or infective) | **Rheumatoid arthritis** |
| Fluid retention (systemic) | **Pregnancy**<br>**Menopause**<br>**Obesity**<br>**Hypothyroidism**<br>**Renal failure** |
| Repetitive movements of the wrist | **Keyboard work**<br>**Vibrating tools** |

(1 mark each for any four of the above).

7) The cyclist has been taking an NSAID for his wrist pain. Inhibition of which gastric enzyme is likely to be the cause of any subsequent chronic gastric pain? (2 marks)

Traditional NSAIDS inhibit both cyclooxygenase 1 and cyclooxygenase 2. Cyclooxygenases' normal physiological role is to convert arachidonic acid into paracrine mediators such as prostaglandins, prostacyclins and thromboxanes. Inhibition of cyclooxygenase 1, COX-1, is significant because constitutive activity of COX-1 is necessary to maintain normal gastric mucosa. The cyclist is likely to be experiencing gastritis because of inhibition of **COX-1**.

(Only one mark for **COX** or **cyclooxygenase**; **COX-1** earns two marks).

8) Two years later the cyclist is diagnosed as having rheumatoid arthritis. What is the underlying aetiology of rheumatoid disease? (4 marks)

Rheumatoid disease is an **autoimmune disease** that occurs after a loss of **tolerance to self** antigens. Specifically the autoimmune attack of **IgM antibodies against the Fc part of IgG** results in the formation and deposition of immune complexes throughout the body. This causes **inflammation** at the sites of deposition.

**Total 20 marks**

## Test Twenty-Four

An 18 year old boy presents to his GP complaining of pain on urination. After a complete history and examination it becomes apparent the young man has been suffering from a urethral discharge for two days.

1) How is he most likely to have acquired this urethritis and what are the two commonest causes? (3 marks)

Urethritis is usually acquired as a **sexually transmitted disease/infection (STI or STD)**. The commonest aetiology of male urethritis is a bacterial infection by either **Chlamydia trachomatis** or **Neisseria gonorrhoeae.**

> The commonest STI is a chlamydia trachomatis infection. In women repeated STI infections increase the risk of infertility. The inflammation and subsequent tissue damage and scarring of the fallopian tubes is believed to be a major cause of the infertility.

2) The patient has a 30 year old female partner, Sarah, who is informed of his condition. Six months later she is diagnosed with squamous cell carcinoma in situ of the cervix (CIN 3). She is distraught and blames her partner's E.coli urethritis for the neoplastic diagnosis. Is Sarah correct? Explain. (4 marks)

**She is not correct.**
**High risk human papillomavirus has been demonstrated to be a direct cause of CIN3** and is most commonly **transmitted through sexual activity**. There is no association between **E.coli infection and CIN3; the E.coli infection is more likely to be related to a urinary tract infection than an STI**. (Neither Chlamydia trachomatis nor Neisseria gonorrhoeae is a direct cause of CIN3. They are associated with CIN3 because they may be present as markers of *sexual activity* which *is* a risk factor for CIN3).

3) Despite treatment she goes on to develop a cervical cancer. Cervical cancers are common in the general population. Name three different cervical malignancies. (3 marks)

Any three of:
**Squamous cell carcinoma, adenocarcinoma, adenosquamous carcinoma, lymphoma, small cell carcinoma and malignant melanoma.**
Secondary malignancies/locally invasive malignancies are also possible answers.

Three months after being diagnosed with cervical cancer and receiving treatment she visits her GP complaining of chest pain. The chest pain was sharp and associated with shortness of breath for two days. There is no radiation, nausea or swelling of ankles. She had been apyrexial.
The chest pain was worsened by the movements of breathing.

4) What is the most likely diagnosis and what drug would you select for initial treatment? (2 marks)

**Acute Pulmonary Embolus**
**Low molecular weight heparin** (commonly enoxaparin, dalteparin or tinzaparin).

The cervical tumour is a risk factor for blood clots. The acute onset of pleuritic chest pain is a classical clinical presentation of a pulmonary embolus (PE). The Well's score accordingly would be at least 4 – equivalent to a moderate risk of PE:

| Wells (PE) Score | |
|---|---|
| **Clinical feature(s)** | **Score** |
| • Clinically suspected DVT. | • 3 |
| • Alternative diagnosis is less likely than a PE. | • 3 |
| • Tachycardia. | • 1.5 |
| • Immobilisation for more than three days or major surgery in the previous four weeks. | • 1.5 |
| • History of DVT or PE. | • 1.5 |
| • Haemoptysis. | • 1 |
| • Malignancy (receiving treatment in the previous six months or palliative). | • 1 |

5) List three possible ECG changes that can occur with the above diagnosis. (3 marks)

Deep S waves in I, Q waves in III, inverted T waves in III (S1,Q3,T3 pattern) (3 marks)
Sinus tachycardia. (1 mark)
Atrial fibrillation. (1 mark)

ST depression. (1 mark)
Right axis deviation. (1 mark)
Right bundle branch block. (1 mark)

The young man with urethritis, Peter, breaks up with his ailing and cancer ridden partner. He is persuaded by his parents to spend some time with his uncle, George, on his farm whilst he sorts out his life. Peter initially does well appearing to adapt to his new location, making friends and finding work at the local newspaper. After about a year Peter's uncle starts to notice some odd behavioural changes. Peter begins to spend more time by himself, he sporadically fails to attend work without any clear reason and pays noticeably less attention to his personal hygiene. His uncle takes him to the local GP who gives Peter a short course of Prozac. However, over the next year Peter inexorably declines, stops going to work altogether and avoids his friends. George finds Peter harder and harder to understand in conversation. Peter claims that to have survived his break up shows that he has god like strength of mind and so he must be a god. On Sunday evenings, their weekly game of chess ends unsatisfactorily with Peter claiming that he has discovered a new way for the rook to move.

6) What is the most likely diagnosis of Peter's condition? (1 mark) Identify four supporting clinical features. (4 marks)

The most likely diagnosis is **schizophrenia**.

Supporting clinical features can be any four of:
Peter is **delusional** in believing that he is a god. (1 mark)
The difficulty in communication with his uncle probably reflects **disorganized speech**. (1 mark)
He shows social and organizational dysfunction in **diminished self care**, **work deterioration** and **avoidance of friends**. (1 mark each, maximum of 3 marks)
His GP did **not appear to find an organic cause** for the behavioural changes. (1 mark)
The chess games may represent a combination of **delusion, disorganized speech and behavioural change**. (1 mark)

**DSM-IV Criteria for Schizophrenia**
A) Two or more symptoms, each present for a significant portion of time during a one month period:

Delusions
Hallucinations
Disorganized speech
Grossly disorganized or catatonic behavior
Negative symptoms (e.g. flattening of affect)

B) Social/occupational dysfunction.

C) Continuous signs of disturbance persist for at least 6 months.

**Total 20 marks**

## Test Twenty-Five

A 30 year old man attends for a consultation stating that six months ago he tried but failed to quit smoking. However, he now wants to try again and has already cut down the number of cigarettes he smokes each day. He wants your help and advice to successfully quit smoking.

1) List and briefly describe four of the ICD-10 diagnostic criteria for substance dependence. (8 marks)

Any four of:

- A strong desire or sense of **compulsion to take the substance.**
- Impaired capacity to control substance-taking behaviour in terms of its onset, termination, or levels of use, as evidenced by the substance being often taken in **larger amounts or over a longer period** than intended, or by a persistent desire or unsuccessful efforts to reduce or control substance use.
- A **physiological withdrawal** state when substance use is reduced or ceased, as evidenced by the characteristic withdrawal syndrome for the substance, or by use of the same (or closely related) substance with the intention of relieving or avoiding withdrawal symptoms.
- **Evidence of tolerance** to the effects of the substance, such that there is a need for significantly increased amounts of the substance to achieve intoxication or the desired effect, or a markedly diminished effect with continued use of the same amount of the substance.
- Preoccupation with substance use, as manifested by important alternative **pleasures or interests being given up** or reduced because of substance use; or a great deal of

time being spent in activities necessary to obtain, take or recover from the effects of

- Persistent substance use **despite clear evidence of harmful consequences** as evidenced by continued use when the individual is actually aware, or may be expected to be aware, of the nature and extent of harm.

2) In the **transtheoretical model of behaviour change**, smokers can be assigned to one of five stages of change. A) At what stage of change is this patient? B) State two items in the question stem that support your answer. (3 marks)

A) **Preparation stage.**
B) i) Recent **previous attempt** to quit smoking.
ii) Has a **plan of action**.
iii) Action is planned in the **near future**.

3) Besides the stage of change identified in the previous question, name four other stages of change. (4 marks)

**Precontemplation**
**Contemplation**
**Action**
**Maintenance**

4) List and briefly describe the "5 As" that many clinicians use to guide their delivery of smoking cessation advice. (5 marks)

**Ask**
Identify and document current tobacco use level for each patient at each visit.
**Advise**
In an unequivocal manner, urge the tobacco user to quit.
**Assess**
Determine whether the smoker is willing to make an attempt to quit at this time.
**Assist**
For the patient willing to make an attempt to stop, use supportive counselling and pharmacotherapy as necessary.
**Arrange**
Organize follow-up contact, in person or by telephone, preferably within the first week after the cessation date.

**Total 20 marks**

# Test Twenty-Six

As a conscientious student you have worked so hard on your undergraduate biochemistry that you start to dream about it in your sleep. In fact in your dreams you are a crime scene investigator, trying to help track down and capture a particularly ruthless and brutal serial killer. Like all good crime scene investigators you have 100% recall of everything you have ever been taught and have a tendency to speak in taut acronym-laden bursts ......

Your serial killer has become overconfident and at the latest crime scene you identify some drops of the killer's blood.

1) What is the name for the simplest classification of blood group antigens? (1 mark)

**Lewis** antigen system. (Or **ABO** system) (1 mark)

2) Your blood tests indicate that the killer has the most rare blood type. Which blood type is this? (1 mark)

**AB** (1 mark)

3) A simple test carried out on the blood displays the chromosomes. What is the name of this display?

**Karyotype** or **karyogram** (1 mark)

4) The test confirms that the serial killer is male. Further tests indicate that your criminal is carrying the genetics (triplet repeat expansions) for a rare neurodegenerative disorder. This disorder is expressed in every generation and its age of onset is earlier and more severe with each generation.

i) In terms of simple Mendelian genetics what type of inheritance is this disease likely to show? (2 marks)
ii) What is the term for the fact that the disease occurs earlier with each generation and is likely to be more severe? (1 mark)
iii) What is this disease called? (1 mark)

i) **Autosomal dominant** (2 marks)
ii) **Anticipation** (1 mark)
iii) **Huntington's disease** (or Huntington's chorea) (1 mark)

You suspect that part of the killer's motivation may be fatalistic appreciation of his own disease and disease prognosis.

On a thorough examination of the genome it is noted that this extremely unfortunate serial killer is a carrier for the sickle cell disease (he is a heterozygote and so is asymptomatic).

5) Based on this information can you speculate regarding the likely ethnic group of the killer? (1 mark)

He is likely to be of **Afro-caribbean** ethnicity/descent. (1 mark)

You pass on this wealth of information to the police, who are enormously grateful and make you their CSI of the week.

6) What is the nature of the DNA disorder in sickle cell disease? (2 marks)

It is a point mutation **substitution** that involves a conversion of an **adenine base to a thymine base**. (2 marks)

7) Considering how damaging the effects of having sickle cell disease are, it is surprising that it has not been eliminated as a result of negative evolutionary pressures. Can you suggest a reason why the sickle cell trait (sickle cell carriers) still exists in the population? (2 marks)

Because it confers a **survival advantage**. In fact it allows the carrier to resist **malarial infections**. (2 marks)

8) Because of this selection pressure, Hardy-Weinberg population genetics do not apply well to sickle cell population genetics. What are the assumptions of Hardy-Weinberg population genetics? (5 marks)

**Population is large.**
**No migration** into or out of the population.
**Random mating.**
**Mutation rate remains constant** (or no mutation at all).
**No selection of alleles** (neither negative not positive).
(1 mark each)

9) State the two Hardy-Weinberg equilibrium equations (or expressions). (3 marks)

**$p + q = 1$** (1 mark)
**$p^2 + 2pq + q^2 = 1$** (2 marks)

**Total 20 marks**

## Test Twenty-Seven

A 57 year old female politician attends her GP surgery to discuss her recovery from a recent myocardial infarction. Because she is a driven and intelligent person she has done some internet research on this issue and as a result she has a lot of questions. You feel under pressure to maintain your professional credibility by answering as clearly and as accurately as you can.........

1) She has noticed that many of the complications of a myocardial infarction involve either arrhythmias or heart failure (including cardiogenic shock). You agree, but feel duty bound to alert her to the wider range of complications. Name five other classical complications of a myocardial infarction. (5 marks)

Any five of:

- Another **myocardial infarction** (or progression of the first myocardial infarction).
- **Angina** or **unstable angina.**
- **Mitral incompetence** (regurgitation).
- **Pericarditis.**
- **Cardiac rupture** (possibly leading to a **cardiac tamponade**).
- **Mural thrombosis.**
- **Ventricular aneurysm.**
- **Dressler's syndrome.**
- **Pulmonary embolus.**
- **Sudden death.**

2) Your patient remembers reading about an autoimmune complication of a myocardial infarction but cannot remember the name. What is the name of this complication? (1 mark)

**Dressler's syndrome.**

> Dressler's syndrome is a delayed complication of a myocardial infarction. It can occur weeks or months after a myocardial infarction.

3) Name two core signs or symptoms of Dressler's syndrome. Briefly indicate the underlying pathogenesis of each. (4 marks)

Any two of:
**Cardiac pain or pericarditis** – autoimmune attack on pericardium and consequent inflammatory pain.
**Pericardial effusion** – inflammatory effusion.
**Pleural pain or pleuritis** – autoimmune attack on pleura leading to inflammation.
**Pleural effusion** – inflammatory effusion (exudate).
**Fever** – significant volume of tissue under autoimmune attack, leading to the action of inflammatory pyrogens.
**Shortness of breath** – impaired cardiac and respiratory function.

4) Name two other autoimmune diseases that directly target the cardiovascular system. (2 marks)

- **Rheumatic fever.** (1 mark)
- **Temporal arteritis** (or any autoimmune vasculitis). (1 mark)

Any of the diseases in the table below would also be acceptable answers:

| **Autoimmune vasculitides** (*singular* vasculitis) include: | |
|---|---|
| Behcet's disease | **Small vessel vasculitis** |
| Buerger's disease | **Small vessel vasculitis** |
| Central nervous system vasculitis (or cerebral vasculitis) | **Small vessel vasculitis** |
| Churg-Strauss disease | **Small vessel vasculitis** |
| Cryoglobulinaemic vasculitis | **Small vessel vasculitis** |
| Henoch-Schonlein purpura | **Small vessel vasculitis** |
| Hypersensitivity vasculitis (or cutaneous leukocytoclastic vasculitis) | **Small vessel vasculitis** |
| Microscopic polyangiitis | **Small vessel vasculitis** |
| Rheumatoid vasculitis | **Small vessel vasculitis** |
| Wegener's granulomatosis | **Small vessel vasculitis** |
| Kawasaki disease | **Medium vessel vasculitis** |
| Polyarteritis nodosa | **Medium vessel vasculitis** |
| Polymyalgia rheumatica | **Large vessel vasculitis** |
| Takayasu's arteritis | **Large vessel vasculitis** |
| Temporal arteritis (or giant cell arteritis or cranial arteritis) | **Large vessel vasculitis** |

5) Indicate two classical long term complications that can occur as a result of any autoimmune disease. (2 marks)

Any two of:

- Increased risk of **another autoimmune disease.** (1 mark)
- Increased **risk of lymphoma**. (1 mark)
- **Treatment effects** – such as glucocorticoids causing **Cushing's syndrome**. (1 mark)

6) The politician has a sister who suffered from a myocardial infarction at 55 years of age. Assuming that this myocardial infarction occurred as the result of a thrombosis, where in the vasculature is this thrombosis most likely to have occurred? (2 marks)

Within the **first two centimetres** of the **left anterior descending artery** (anterior interventricular artery). (2 marks)

> Approximately 50% of myocardial infarctions occur with a thrombosis in the proximal part of the left anterior descending (LAD) artery.

7) Assuming that this myocardial infarction caused ST elevation, which ECG leads are most likely to reveal this change? (2 marks)

The acute coronary syndrome occurring because of a thrombosis in the LAD is most likely to cause changes in the *anterior* chest leads; **V1, V2, V3 or V4**. (0.5 mark each)

8) Percutaneous coronary intervention was not successful and so thrombolysis was used in the management of this myocardial infarction. Name two such commonly used fibrinolytic agents. (2 marks)

Any two of:
**Streptokinase**
**Tissue plasminogen activator (tPA)/alteplase**
**Urokinase**
**Reteplase**
**Tenecteplase**
(1 mark each)

**Total 20 marks**

## Test Twenty-Eight

A 27 year old bicycle courier presents to you as an emergency admission. He complains of leg pain after being driven into by a Lexus. On examination his lower leg is lacerated, swollen and deformed with loss of the ability to weight bear. The courier is in a great deal of pain and you immediately prescribe oral analgesics and request X-rays of the leg. The X-rays show a mid tibia-fibula fracture. The leg is placed in a cast.

1) You suspect that the muscles of the lateral compartment have been crushed and their function impaired. Name the two muscles of the lateral compartment of the lower leg. (2 marks)

**Peroneus (fibularis) longus.**
**Peroneus (fibularis) brevis.**

2) What are the normal actions of theses muscles? (2 marks)

**Plantarflexion.**
**Eversion of the foot.**

Three days later the patient returns to hospital complaining of exquisite tenderness of that leg. The pain keeps him awake at night. Because the pain is out of proportion with the presumed diagnosis acute compartment syndrome is suspected.

3) What are the generally accepted 5Ps of compartment syndrome? (5 marks)

**Pain.**
**Pallor.**
**Paraesthesia.**
**Paralysis.**
**Pulselessness.**

> The cardinal feature of compartment syndrome is the increase in pressure in the volume affected. This can compromise venous outflow (low pressure) and then compromise arterial inflow (high pressure). Naturally this will cause a loss of pulses. However, the ischaemia and subsequent necrosis will also damage nerve cells as well as muscle cells. Hence this damage to nervous and muscular tissue can be permanent if the increase in pressure is not relieved promptly.

4) A senior doctor mentions that there is a sixth P that can also be useful in the diagnosis of compartment syndrome. Explain to what this sixth P may refer. (2 marks)

The poor blood flow to the compartment means that **thermoregulation becomes difficult**. Hence the affected compartment is usually cooler than the normal compartment on the other limb (contralateral compartment). This is the ***poikilothermia*** effect.

5) As well as the muscle weakness and continuing muscle pain the courier mentions that he also has dark urine. As a result there is a suspicion of impairment of which major intrabdominal organ(s)? (1 mark)

**The kidneys.**
(0.5 mark for liver e.g. haemolytic jaundice)

6) Initially which blood tests would you request to confirm the clinical suspicion? (3 marks)

Renal function should be investigated by taking a venous blood sample to determine:
**Urea** concentration.
**Creatinine** concentration.
**eGFR** (estimated Glomerular Filtration Rate).
**Electrolyte** concentrations – changes of AKI; potassium, sodium, hydrogen, phosphate and calcium ions.
**Myoglobin** concentration.
**Creatine kinase** concentration.
(0.5 mark each)

> Acute compartment syndrome caused by a crush injury to a limb can cause a large release of muscle proteins, including myoglobin that can lead to acute renal failure (acute kidney impairment). The excess myoglobin can precipitate in the renal tubules to cause obstruction. It can also directly colour the urine.

7) A blockage can occur in any of the tubular components of the nephron. Name the tubule components of the nephron through which urine is formed and passes. (5 marks)

**Proximal convoluted tubule.**
**Ascending loop of Henle.**
**Distal convoluted tubule.**
**Descending loop of Henle.**
**Collecting duct.**

**Total 20 marks**

## Test Twenty-Nine

A previously well 15 year old boy presents with abdominal pain that is central abdominal, 7/10 in intensity with a 24 hour history. On examination he has no abdominal scars. Two hours after being admitted into hospital he complains that the pain is worse and is now in the right lower quadrant. He spikes a fever of 38.5°C.

1) What is the most likely diagnosis? (1 mark)
**Acute appendicitis** (no marks for just appendicitis, *acute appendicitis* is required for the mark).

2) Name three endogenous chemical factors that may be causing the pyrexia. (3 marks)

The acute inflammation will cause release of factors that cause pyrexia; classically these are pyrogens such as **IL-1, IL-2, IL-6, TNF-alpha** or **prostaglandins.**

3) What is the definitive management of acute appendicitis? (2 marks)

**Appendectomy** with **antibiotic** cover/supportive management.

4) Describe the most likely macroscopic appearance of the excised appendix. (3 marks)

Any three of:
1) **Enlarged**. 2) **Congested**/erythematous. 3) Pus covered/**Purulent exudate**.
4) **Ruptured**. 5) Possible fecal matter/**fecalith** at cut surface. (1 mark each)

5) On examining the inflamed appendix the histopathologist notes an infiltration of numerous cells with multilobed nuclei. What is this dominant cell type likely to be and is the preponderance of this cell type consistent with your favoured diagnosis? (2 marks)

**Neutrophils** (not just polymorphs or polymorphonuclear cells). Their presence is **consistent with acute appendicitis**.

6) Unfortunately for the patient, on thorough examination of the appendix the histopathologist finds an area that amounts to a carcinoma. What are the key cellular features of malignant cells? (4 marks)

**Nuclear pleomorphism** (variation in shape and size of the nuclei).
**Cellular pleomorphism**.
Increased **nuclear to cytoplasmic ratio**.
Increased numbers of **normal mitoses**.
Increased numbers of **abnormal mitoses**.
Prominent **nucleoli**.
**Hyperchromatism**.

**Invasion** through the basement membrane into adjacent tissue. (0.5 mark for each correct answer)

7) It takes the teenager a surprisingly long time to recover from the surgery. The surgeon notes that the wound takes longer than usual to heal. Name five **general** systemic factors that can decrease the rate of wound healing. (5)

Any five of:
A) Increasing **age**.
B) **Chronic disease** (e.g. diabetes mellitus, thyroid disease).
C) **Medications** (e.g. glucocorticoids).
D) **Poor cardiovascular status** (e.g. peripheral vascular disease).
E) **Dietary deficiencies** (e.g. vitamin C).
F) **Systemic infection**.
G) **Immunodeficiency** or **immunosuppression**.

Only ½ a mark for the answer "*Stress*."
(Local factors excluded from this answer are: Site, size, tissue type, apposition/fixation, infection, local blood supply, foreign material, radiation damage).

**Total 20 marks**

## Test Thirty

1) As a junior doctor working for surgeons, you have to manage many postoperative patients. It is not unusual for one of your patients to be pyrexial.
Name four classic (diagnostic) causes of postoperative fever. (4 marks)

**Table 1: Causes of Postoperative Fever – 5Ws**

| **W**ater | **W**ind | **W**ound | **W**alking | **W**onderdrug |
|---|---|---|---|---|
| Urinary Tract Infection | Pneumonia | Wound Infection | Deep venous thrombosis and pulmonary embolism | Penicillin (beta-lactam antibiotics, isoniazid etc) |

2) What is the first line drug used in the treatment of a urinary tract infection (UTI)? (1 mark)
**Trimethoprim (or nitrofurantoin).**

3) What are the first line drugs used in the treatment of deep venous thrombosis (DVT)? (2 marks)

Low molecular weight **heparin**
**Warfarin**

4) Name the tests used to show that heparin is within its therapeutic range and that warfarin is within its therapeutic range. (2 marks)

Warfarin => Measure **INR** (International Normalized Ratio)
Heparin => Measure **aPTT** (Activated Partial Thromboplastin Time)

5) A 55 year old patient has undergone a heart valve replacement with an artificial valve. As a result she is required to maintain lifelong anticoagulation with warfarin. The aetiology of her heart valve dysfunction is believed to be childhood rheumatic fever. For the first year after the operation the warfarin mediated control of INR is routine. However, after the second year progressively larger doses of warfarin are required to maintain the target INR. During the third year the patient's renal function and cardiac function remain good, however, the hepatic function deteriorates insidiously. The patient reported that she was divorced by her husband shortly after the valve transplant and is currently dating again. When she puts down her handbag the physician hears the clinking of glass bottles.

What is the most likely explanation for deterioration in the control of her INR? (3 marks)

A very common cause of liver impairment in the Western world is **alcohol abuse**. Alcohol induces cytochrome P450 enzymes in the liver that can lead to an increased clearance of medicines such as benzodiazepines, phenytoin and **warfarin**. Larger doses will thus be required to maintain the target INR.

6) If the warfarin levels are maintained at too high a level, to what types of complications will the patient be prone? (1 mark)

**Haemorrhagic** disease (e.g. bruising/haemorrhagic stroke/haemarthrosis).

7) Over the course of the next five years, she develops spider naevi, caput medusae and haemorrhoids. Blood tests show markedly raised AST, ALT and ALP. Concurrently her personal and social life leave her unhappy and she admits that drinking excessive alcohol is a problem. What affective disorder is chronic alcohol abuse most likely to cause? (1 mark)

**Depression**.

8) The GP is concerned about the progression of her psychiatric disorder and decides that she needs pharmacological support. He considers using an SSRI, SNRI or MAOI

medications.
Give an example of each of these classes and briefly describe the primary mechanism of action. (6 marks)

**ssRI** - e.g. **Fluoxetine** – inhibits serotonin re-uptake in the CNS.
**snRI** - e.g. **Venlafaxine** – inhibits serotonin and noradrenaline re-uptake in the CNS.
**MAOI** - e.g. **Selegiline** – inhibits breakdown of catecholamines in the CNS.

(2 marks each)

This question is concerned with the different types of antidepressants. Fluoxetine is commonly called Prozac.

**Total 20 marks**

Printed in Great Britain
by Amazon